SCIENCE AND TECHNOLOGY EDUCATION
AND FUTURE HUMAN NEEDS

Volume 3

GW00725477

Education,
Industry and Technology

Science and Technology Education and Future Human Needs

General Editor: JOHN LEWIS *Malvern College, United Kingdom*

Related Pergamon Journal

INTERNATIONAL JOURNAL OF EDUCATIONAL
DEVELOPMENT

Editor: PHILIP TAYLOR

Throughout the world educational developments are taking place: developments in literacy, programmes in vocational education, in curriculum and teaching, in the economics of education and in educational administration.

It is the purpose of the International Journal of Educational Development to bring these developments to the attention of professionals in the field of education, with particular focus upon issues and problems of concern to those in the Third World. Concrete information of interest to planners, practitioners and researchers, is presented in the form of articles, case studies and research reports.

Education, Industry and Technology

Edited by

D. J. WADDINGTON
University of York, United Kingdom

Assisted by: J. W. Steward with
R. T. Allsop, M. H. Gardner,
J. S. Holman, D. McCormick
and S. Ware

Published for the

ICSU PRESS

by

PERGAMON PRESS
OXFORD · NEW YORK · BEIJING · FRANKFURT
SÃO PAULO · SYDNEY · TOKYO · TORONTO

U.K.	Pergamon Press, Headington Hill Hall, Oxford OX3 0BW, England
U.S.A.	Pergamon Press, Maxwell House, Fairview Park, Elmsford, New York 10523, U.S.A.
PEOPLE'S REPUBLIC OF CHINA	Pergamon Press, Room 4037, Qianmen Hotel, Beijing, People's Republic of China
FEDERAL REPUBLIC OF GERMANY	Pergamon Press, Hammerweg 6, D-6242 Kronberg, Federal Republic of Germany
BRAZIL	Pergamon Editora, Rua Eça de Queiros, 346, CEP 04011, Paraiso, São Paulo, Brazil
AUSTRALIA	Pergamon Press Australia, P.O. Box 544, Potts Point, N.S.W. 2011, Australia
JAPAN	Pergamon Press, 8th Floor, Matsuoka Central Building, 1-7-1 Nishishinjuku, Shinjuku-ku, Tokyo 160, Japan
CANADA	Pergamon Press Canada, Suite No. 271, 253 College Street, Toronto, Ontario, Canada M5T 1R5

First edition 1987

Library of Congress Cataloging-in-Publication Data
Education, industry, and technology.
(Science and technology education and future
human needs; vol. 3)
Includes bibliographies.
1. Science—Study and teaching—Congresses,
2. Science—Study and teaching—Developing
countries—Congresses. 3. Curriculum planning—
Developing countries—Congresses. 4. Industry—
Congresses. 5. Technology—Developing countries—
Congresses. I. Waddington, D. J. II. Series.
Q181.E47   1987   507   87-2344

British Library Cataloguing in Publication Data
Waddington, D. J.
Education, industry and technology.—
(Science and technology education and
future human needs; v. 3).
1. Industry — Study and teaching
2. Industry and education
I. Title.   II. Series
338'.007   LC1081
ISBN 0-08-033913-1 (Hardcover)
ISBN 0-08-033914-X (Flexicover)

Printed in Great Britain by A. Wheaton & Co. Ltd., Exeter

Foreword

The Bangalore Conference on "Science and Technology Education and Future Human Needs" was the result of extensive work over several years by the Committee on the Teaching of Science of the International Council of Scientific Unions. The Committee received considerable support from Unesco and the United Nations University, as well as a number of generous funding agencies.

Educational conferences have often concentrated on particular disciplines. The starting point at this Conference was those topics already identified as the most significant for development, namely Health; Food and Agriculture; Energy; Land, Water and Mineral Resources; Industry and Technology; the Environment; Information Transfer. Teams worked on each of these, examining the implications for education at all levels (primary, secondary, tertiary, adult and community education). The emphasis was on identifying techniques and resource material to give practical help to teachers in all countries in order to raise standards of education in those topics essential for development. As well as the topics listed above, there is also one concerned with the educational aspects of Ethics and Social Responsibility. The outcome of the Conference is this series of books, which can be used for follow-up meetings in each of the regions of the world and which can provide the basis for further development.

JOHN L. LEWIS
Secretary, ICSU-CTS

Acknowledgements

To discuss such a wide range of topics effectively in a single Workshop — and particularly one of such a participatory nature — needs much work in both planning and during the Workshop itself. To make it truly international adds another dimension.

Any success we may have had is due to all those who so willingly gave their time in the two years before the Workshop, in particular Terry Allsop, Jack Holbrook, John Holman, O. C. Nwana and Joseph Yakubu. We are also indebted to my co-leader, Professor M. H. Shrinivasan and to Dr Pravas Mahaptra and Dr R. J. Narahari, all of whom led double, if not triple, lives during the Conference being heavily involved in their own work at the Indian Institute of Science, in taking part wholeheartedly in our activities and, to use cricketing parlance, in fielding supremely well all the awkward balls that came to them, in providing facilities and help to all of us who were visiting their city.

I am most grateful to John Steward, Terry Allsop, Marge Gardner, John Holman, Dave McCormick and Sylvia Ware who gave their time during and after the Conference to help to edit the book.

Finally, my thanks go to John Lewis without whose inspiration this Conference would not have occurred.

Bangalore, Rio de Janeiro, Tokyo and York   D. J. WADDINGTON
September 1985

Permissions

We are grateful for permission to reproduce the following material within this book:

The HSC Physical Science Examination papers are quoted in Chapters 36 and 37 with permission from the Victorian Institute of Secondary Education, Melbourne, Australia.

The question reproduced in Chapter 37 from the course 'Science for All' is used with permission from the author, Terry Hitchings, and those from the SATIS course with permission from the Project Organiser, John Holman.

Contents

Section C: Technology in the secondary school curriculum

Section D: Industrial and technological issues: examples of some secondary science curricula

Section E: Industrial and technological issues: some teaching strategies in the secondary science curricula

Section F: Industrial and technological issues in the secondary science curriculum: Assessment

Section G: Education and the world of work

Section H: The role of tertiary institutions in development

Section I: Technical training for development

Section J: Making curricula relevant for industry: the role of teacher training

Section K: Co-operative education

Introduction

As with so many educational issues being discussed worldwide, the relationship of industry and technology within primary, secondary and tertiary curricula exercises the minds of teachers and others involved in education — in the so-called developed and developing countries alike. Often the problems are similar; it is the degree to which the problem affects a nation's prosperity and development that differs so markedly.

The specific issues may be different at different levels of education. For example, in many developing countries, the majority of children will only be able to obtain primary school education and if it is a school in a remote area or in a deprived urban environment, the standard of teaching and resources may be very low. Yet to break down prejudices, to open minds of pupils and to enable the next generation to be better equipped to help in the development of their country, the introduction of a *suitable* technological bias to the curriculum will be all-important. This is one of the most important problems facing us.

On turning to secondary schools, these and other matters must be considered. Science is taught in secondary schools to students who have many different careers in mind. Schools not only prepare students for scientific and technological studies at tertiary level but they must also try to provide a science education for other aspiring tertiary level students; they are also involved in preparing students to become technicians and related workers. We must also satisfy the needs of those whose formal education stops when they leave school to enter an adult world in which both work and leisure is profoundly affected by science. Thus, we must provide a scientific education for all students, whatever their career choice is and whatever their aptitude.

Different countries have different priorities. Some consider first and foremost the production of scientists and technologists; others are simply trying to provide enough teachers to continue a school science programme; others have the comparative luxury to be able to consider providing a general scientific education for all. The country's political and economic system set the boundary conditions.

Science was introduced into schools as an academic discipline. In the 1950s and 1960s curricula in many countries were altered and it was common to hear the phrase "education through science". In particular, the content changes were directed to stressing the principles of the subject. Much of the new methodology has been concerned to make the pupils participate actively in the learning process, both by a more determined effort to include experimental work as an integral part of the curriculum and also by making the theory lessons more pupil-centred.

This trend is being followed by many countries in the early 80s, but where curricula have been developed earlier one can see a counter-current. Doubt is being expressed about both content and methodology. In content, the so-called "conceptual approach" is heavily criticised. Further, there is an increasing awareness too that arranging effective pupil-centred work is often too demanding in time and skill for many teachers. Moreover, it is felt by some critics that the content of courses has been controlled or heavily influenced by the professional scientist and that the curriculum is being concerned with producing future scientists.

Writing of the curricula developed in the last 20 years in which he has played an important role, John Lewis writes,[1]

> . . . in one aspect these innovations (the curricula of the 60s and 70s) have failed. The new programmes have tended to look inwards on themselves. In searching for patterns to establish, say, the Newtonian laws of motion, the relevance of these fundamental laws to the world in which the pupils live have been lost. . . . There is a tendency for these laws to be seen as an end in themselves. Though the programmes have shown something of how the scientist thinks, they have done little to show the relevance of that science to the world outside the laboratory.

The relevance of "that science" to the world outside the laboratory is a key to the Conference in Bangalore. There are several different, yet interrelated strands in this discussion, when considering secondary science curricula and its relation to industry and technology. One is that we should make sure that the science curriculum is relevant to and should inspire future scientists and technologists. There is no such thing as "European science" or "Indian science". The principles are the same. It is the way these principles are used which may differ. Further, we must ensure that all students are able to appreciate the consequences of scientific development, for good and for evil. Science teachers must take more responsibility for this.

Here is a dilemma: although it may seem attractive to introduce into the curriculum some teaching that relates science to society, science to industry and technology, there are divergent views on how this is to be achieved. Some will argue that the essence of science is as it is largely taught today — to study simple situations with, if possible, a single variable: in this way, cause can be related to effect and natural laws formulated. But to study the role of science in society is in a sense to go to the opposite extreme; to study highly complex situations governed by a large number of parameters, many of them not identifiable. Moreover, it must be remembered that, at the secondary school level, students have relatively little "hard" science to enable them to formulate their ideas in relation to society. Possible answers to this dilemma have been proposed in several countries and are discussed in detail in this book.

There is a similar width of interests in tertiary level teaching and professional education. The phrase "Education, Industry and Technology" covers courses for the professional scientists and engineers as well as courses for other non-science based students. In developed and developing countries, science and engineering courses have often been constructed so that future research students can be produced. These courses have become too inward-looking. This is a particularly tragic policy in developing countries where the facilities for modern research are often lacking and the minds of some of the best students are effectively sterilised. The students become frustrated; employers despair; the country loses some of its best potential. Worse follows: the training given to research students is often inadequate in terms of the economic development of the country. Those students then go to developed countries for post-doctoral training and again receive inappropriate training, inappropriate to the industry and technology of their country.

We must also be concerned with the supply of technicians. It is too easy for scientists and technologists in developed countries to forget the important role technicians play in development. One has to only spend a few moments in institutions in some developing countries to see what happens when there is not adequate technical help. Equipment, bought with the little hard currency available, lies idle for lack of maintenance — even at a relatively simple level.[2] Other equipment which with some forethought could be made in the country, is nevertheless obtained from developed countries, again at great cost. There are probably enough scientists and engineers in many countries. They are unable to work effectively for lack of support.

These are some of the problems which face our educational systems. They have been discussed and described many times. In Bangalore we accepted that these problems exist; our concern was *how* to solve them by learning from each other's experiences — and see whether the solutions sought and perhaps found elsewhere can provide us with hope.

We felt that the *methods* of teaching about industrial and technological issues are just as important as the *content* of what is being taught. Traditional methods such as teacher instruction and reading texts certainly have their place, but are unlikely to be very effective on their own. Students need to be positively involved in *activities* related to those issues. We therefore planned a number of workshops in Bangalore in which participants took part in activities such as playing simulation games, planning industrial visits, technological problem-solving and the evaluation of resources. We wanted to ensure that the ideas generated in these activities were taken home and implemented by the participants.

As a result of the work at the conference in Bangalore, the topics we have chosen for inclusion in this book which is the outcome of our labours are:

— Industrial and technological issues in primary school science
— Building applications into secondary science curricula
— Strategies for teaching, including the use of games and simulations, industrial visits, work experience programmes, and ways to promote technology as the means for solving problems
— The needs of industry (looking at education from the perspective of industry) and the role of tertiary institutions in development

References

1. Lewis, J. L. *General Introduction*, Science in Society Project, Association for Science Education, Hatfield, 1979.
2. Sane, K. V. *Science Education through Self-reliance*, The Commonwealth Foundation, London, 1984.

Section A
Industry, Technology and the Primary School

Introduction

In the first paper, Harlen discusses in realistic terms what is appropriate, based on experience of such work with primary school children, when studying industry and technology at this level. At the end of her paper, she raises some essential points.

(1) Is there a distinction between science and technology at this level?
(2) What are the educational values in activities relating to science and technology in primary schools?
(3) How can continuity and progression in the curriculum be achieved?
(4) What criteria should be applied in the selection of activities?

One possible way of involving industry and technology in the primary school curriculum is to make a visit to an industry. Tanuputra gives a specific example from work done in Indonesia. His account should be compared with the discussion of the visit to Bangalore industries undertaken by participants at the Conference (page 185). Industrial visits take the child to the technology. It is also possible to bring the technology to the child in the classroom and such an experience is described by Violino in his chapter.

During the workshop at the Bangalore Conference small groups considered the value of a visit to an industry in terms of the learning which would take place in it. Examples of two of the outcomes are given below.

Visit to a textile dyeing factory

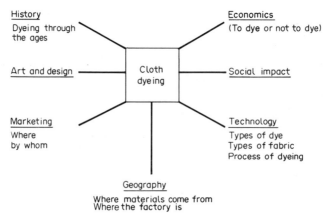

History
Dyeing through the ages

Economics
(To dye or not to dye)

Art and design

Cloth dyeing

Social impact

Marketing
Where
by whom

Technology
Types of dye
Types of fabric
Process of dyeing

Geography
Where materials come from
Where the factory is

In addition, certain skills should be engendered and certain attitudes to be fostered to the maximum extent possible:

Attitudes	*Skills*
Appreciation	Observing
Co-operation	Interpreting
Self-reliance	Comparing
	Pattern finding
	Communicating
	Measuring

Visit to a bakery (10–11-year-olds)

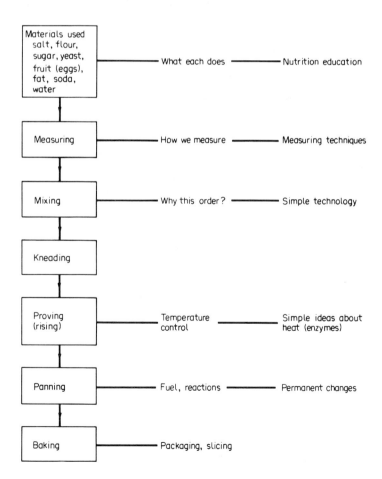

Other issues: the way people work, the effects of mechanisation.

Prior to follow-up work: comparison with home-made bread, do some baking, experiments with yeast, history of bread making.

1

Industry and Technology in the Primary School: Some General Principles

W. HARLEN
University of Liverpool, UK

There are few places in the world where industry does not impinge on children's lives. Anyone who eats packaged food, wears clothes bought from a shop or market, uses energy in the form of electricity, gas or oil, owns or uses a vehicle, does so as the result of the existence of industry and the use of technology. These facts alone argue for the study of industry and technology in an approach to education which aims to help children understand the world around them. The case becomes even stronger if we add the point that most children will find employment in industry when they leave school, which, for a large proportion of children in many developing countries, is directly after leaving the primary school. Postponing any study of the world of work to the secondary school is therefore too late; the children who need it most would be the ones to miss it.

However, arguing in general terms that such study would be useful to future citizens is some way from showing that it is a worthwhile activity in school at the primary level. To make this case we need to examine more carefully the kinds of experiences that may be required and to consider how appropriate and worthwhile they are for young children. The age, maturity and learning characteristics of the children have to be considered in designing this element of the curriculum.

Having argued a general case for the relevance of some study of industry and technology in primary education, we should consider the aims of such study in more detail. What would we want children to derive from such work? A possible answer is summarised in the following six points:

1. Information. Most primary children's knowledge about industry is constrained by the experience of their parents' work and the industry in their locality. A wider view can be given. A teacher who enquired about children's ideas of industry found that, before any work on the subject was started, the children equated "industry" with "manufacturing". Even though the school environment was rural, none of the children regarded agriculture as an industry. The words they used to describe industry were such as "noisy", "dirty", "busy", "dangerous" and girls were particularly fearful of the thought of working in these places. Later, after visits and follow up work, the children had a more balanced view of what industry included and had begun to look behind the external features to see that satisfying and important work was being done. A central aim is thus to give children enough information to counter their assumptions and prejudices and to replace these by knowledge of the range of industry, of its link with their lives and of its use of technology.

2. *Cognitive skill development.* In studying industrial processes and technology there is opportunity for development in the recognition of the sequence in events, in representation through models, charts and flow diagrams, in problem solving and in higher level skills that are involved in, for instance, critical path analysis used in designing production processes. As will become clear through the examples given later, the important experience for this learning comes from being involved in technology and in making sense of production processes rather than simply *learning about* these things.

3. *Attitudes.* Associated with the ideas children often have, mentioned in 1. above, are attitudes which tend to be negative, arising from bewilderment, dislike of noise and dirt and the impression that people in factories are no more than operators of machines. For girls, it is a man's world in which they do not see themselves as having a role. A related attitude comes from equating "work" with "physical effort" and gives rise to a negative view of managers and office workers who may be seen as unnecessary, even lazy. Clearly such attitudes are unhelpful in understanding the co-operative nature of industry and appreciating the variety of roles that have to be taken. Experience is that these attitudes are readily changed when children become involved in studying industry especially if they take part in the production process themselves.

4. *Social learning.* Visiting industry, or inviting workers into the school, enables children to see the workers as people not as operators who are as faceless as machines. In conversation about their work the children can come to realise that most jobs, however routine, have their challenge and can bring satisfaction. It also becomes evident that the workers fit into a scheme of things and someone has to direct what they do. The social hierarchy in an industrial concern begins to emerge and children may be able to appreciate that the power structure is related to responsibility for planning and is not a matter of empty status. It is as important to the understanding of industry to begin to sort out the social relationships, particularly between workers and managers, as it is to know about the physical operations which are carried out.

5. *Science-related learning.* Although much that has already been said

relates to learning in science, it is perhaps worth underlining these contributions a little more. Much of what children may learn in the classroom — about materials, forces, the effect of heat, techniques of measurement — begins to mean something in real terms when they see it in operation in a production process. In some cases the theory may be introduced after a visit which raises questions about why certain materials, techniques and devices are used. For example, a visit to a factory which makes kettles led to investigation of different metals and their suitability for use in making kettles (rusting, denting, expansion, soldering etc.), to the investigation of a bi-metal strip used in the automatic switch, to the exploration of the stability of different shapes and a survey of people's preferences for shape and colour.[1]

6. *Learning related to other subjects.* In the work about kettles just mentioned, not surprisingly, a considerable amount of mathematics emerged — measurement of the capacity of different kettles, of the area of metal required for making a kettle, of the cost of a cup of tea, and so on. Other subjects lend themselves to geographical studies (why the industry is sited where it is), historical studies (how demands for different commodities have changed, how transport and marketing have developed and the nature of working conditions past and present) and religious studies, even. There is scope for art and craft in model making and the solution of technological problems and of course in all parts of the work the potential for language development is considerable. When children are involved in finding out about industry they may have to write letters, descriptive accounts and learn to make notes; they also have opportunity to exercise spoken language in new settings, when interviewing people, questioning strangers, debating issues.

Feasibility and relevance for young children

The above points concerning the potential values of learning about industry and technology may seem persuasive in theory, but what can be done in practice? Is it feasible to involve young children in studying what are often complex issues and organisations. What limits are imposed by the immaturity of the children in the primary school?

The answers to all these questions are interconnected. There is a great deal which can be done in practice — and it is the purpose of this paper to explore some of it — but it has to be planned taking into account the nature of the intellectual, social and physical capabilities of the children. It is encouraging, however, that those who have embarked on this work with primary children invariably report surprise at just what can be done. For example, when the secondary school based Fulmer Industry Project first extended its work on introducing an awareness of technology into the English primary school, the project team made the following comment:[2]

"Our reluctance to introduce technological activities into the primary schools was largely based on ignorance of how primary schools operated. The eventual successful outcome of the project is largely due to the contrast between teaching methods and attitudes found in the primary schools as compared with those adopted in the secondary system. Surprisingly, the primary system was found to be more in tune with the real world than we had seen in many parts of the secondary system; quite wrongly we had assumed the reverse situation would have been the case. Although this finding will not be any surprise to many teachers, people outside education would not have expected that in many instances courses become more artificial as one proceeds up the educational ladder.

We were delighted to find that primary schools operate in a climate not too dissimilar to that found in an engineering design office; for, in the primary school, children are encouraged to work in groups on open-ended problems, to cross subject boundaries and to let their interest, inquisitiveness and enthusiasm motivate them."

For such satisfying work, however, the project has to be carefully selected and planned. A visit to an industry is an excellent way to embark on a project with young children, whose thinking gains from concrete experience rather than from theoretical notions. The teacher's work begins long before the project starts in the classroom, however. Once a suitable place to visit has been decided the teacher has to begin a correspondence to negotiate a visit and to make the necessary arrangements within the school — since it is likely to be necessary to take extra teachers or parents to help in supervising the children. A prior visit by the teacher to the industry or place of work is an essential part of the preparation. This is an opportunity not only for the teacher to find out what the children will experience but to tell anyone in the place to be visited who may be acting as guide, demonstrator or informer, something about the children. In some cases those involved may have little idea of the level of background knowledge, length of attention span and interests of primary children. The teacher can help them with their preparation by suggesting some of the questions the children might ask and the sorts of things that will need to be explained in simple terms.

The following guidelines which have emerged from work in industry with children aged 8 to 13,[1] are useful to have in mind in selecting and preparing a visit:

(i) It should not be an isolated event, but the centrepiece of extended work which begins beforehand and continues afterwards.

(ii) The type of industry visited should preferably be one producing something which children recognise and relate to; for example bicycles, toys or domestic utensils. It was found that complex processes involving huge machinery, often very noisy, did not make for successful visits.

(iii) The best size of group to take round a work place has been found to be six to eight children, so to break a class down into groups of this size requires the help of parents or people from the industry being visited.

(iv) When children are issued with worksheets, either by the school or the place being visited, it often happens that these narrow the observations of the children. One report was that "many children were observed spending most of their time buried in their notebooks (or even worse, other children's notebooks!) rather than listening to and observing the operation of the factory or work place". An approach which appears to be a useful compromise between too much and too little prior instruction is to indicate to children beforehand the main areas of interest and enquiry and to leave the children scope to frame their own questions. The value of this approach for process skill development is clear, especially if some preparation time is spent on the development of questioning and observational skills.

(v) In many cases it is useful follow-up work for the children to create a model of the place visited. As well as the problems of scale and construction to be overcome, the activity helps children to realise the sequence of events. For older children flow diagrams would have similar benefits.

(vi) Inviting someone from the industry back to the school to talk to the children some time after the visit gives them opportunity to pose questions which have occurred later, during reflection on the visit. The double benefit here is that the adult from the industry has a chance to learn about the children's work and the school environment, which can improve school-industry communication for the good of later visiting groups.

A story from Wales

At the time of the project work, this small school in a rural area of the county of Clwyd in Wales had 57 children aged 5–14+. As the school had only three full-time teachers the children were grouped in classes with quite a wide age range. A further complication was the use of two languages in the school — instruction for some of the children was in Welsh and for others in English. In order to give the children some insight into the relevance for work outside school of their studies in school, a project was planned round visits to a variety of industries. One of these, and the follow up work, was described by the teacher as follows:

The most successful visit was to the rural workshop, arranged to contrast with the previous ones; this industry employed just two people. The children were shown each stage in the production of a soft toy — a "sophisticated frog". They were then allowed to participate in the making of a frog — cutting, stuffing, making eyes, sewing and printing. A good deal of interest was shown in the use of the home-made tools; these were unique, and could only do the tasks required by this one business. They served as important examples for the children of how any industry has to cope with its problems, and attempt to make work easier for its employees.

The trip culminated in the children being presented with the frog that they had made. They chose a name for it — Madog — and this was printed on the case that the frog carried around its neck. After returning to school, the trip was the constant discussion point, and Madog was continuously being pulled out of its store cupboard for display. The children decided that they wished to produce something in a similar manner, and the teachers suggested that they think of something that could be sold at the school's Christmas fair. The children quickly came up with the idea of using old Christmas cards but the end product was not decided. A tape recorder was left running to eavesdrop on the planning meetings. As we see, the discussion soon comes around to the decisions that most new businesses have to make.

"Where can we get enough cards? . . . ask people."

"What do we need? . . . scissors that go in and out . . . tag things . . . a thing for making holes."

"What part of the cards do we need? . . . what part can we cut out? . . . depends on the quality of the cards . . . some are floppy . . . make sure there's no writing on the back of the picture . . . can we put glitter on them? — so we would have to buy it . . . we need to make a profit . . . they are easy to make."

A long discussion followed about whether they needed a "manager" and whether they should put "Made in . . . . . " on the product.

"How much can we charge? . . . If the price is too high people won't buy them . . . if we sell them cheaper a lot of people will buy them and we will make the same profit, but we must get rid of them all."

"If we use old cards to make new cards we need to write rhymes . . . I'm no good at them . . . Leave writing out and put 'To' and 'From' on them . . . Yes, then they are like gift tags."

And so it was decided to use old Christmas cards, and make gift tags from them. In order to get enough old cards, the group had to write a suitable letter, duplicate it on the school's spirit duplicating machine, and then distribute the letter to each child in the school. As a result of the children's request, sufficient old cards were received to enable the "small industry" to get off the ground. Before any work was started, the group discussed how the jobs were going to be tackled — again a tape-recorder eavesdropped on their conversation.

The group's conversation centred on the distribution of jobs, and how they were going to be carried out. The children decided on a "production line": each child had a task to complete before passing the work on to the next in the line. They realized that some jobs were going to take longer than others, and therefore two people had to be allocated to this work. Jobs were distributed as follows:

Marking of cards for cutting — all the group

Cutting — two children

Punching hole — one child

Tying thread — one child

Packing — all of the group after their work had been finished

While working in this way, the children realized how monotonous working on a "production line" can be and they were soon changing over jobs, and deciding to move along the line of work every ten minutes. As the evening of the Christmas sale approached, many cards were still unprocessed and so other children were "employed". It was interesting to note that the newcomers to the group were given the most difficult or tiring jobs, for example, cutting with large pinking shears. The original group were asked if this procedure was fair, and they agreed that these were jobs they disliked, but as they were the originators of the business the "workers" had to do as they were told. (If there had been time, the new "workers" could have had interviews, and their job descriptions given to them — it would have been very interesting to hear what kind of questions would have been asked.)

As the work was nearing completion, the original group started to make posters to put up on their stall in the sale. They also drew up a duty rota for serving on the stall, so that they all had an opportunity to leave and go around the other attractions in the sale. The sale of gift tags raised nearly £5, and the children are now deciding how they can spend this money to buy something for the school. Some work on costing and profit could have been carried out if the project involved buying any of the materials, but in this case all the old cards and thread were supplied free by parents, and all the money received was profit, and thus there was little opportunity to investigate this concept of the work.

R. Garem Jones, quoted in.[1]

In reflecting on this project the teacher pointed out the obvious difficulties of carrying it out in a small rural school — the few teachers available, the problems of the necessary preliminary preparation. Despite the problems the effort was felt most worthwhile. The first experience of taking children on such visits showed that much more work than expected was in fact generated. Thus it was not desirable to visit more than one place in a term (and so no more than three at most in a year) and this reduced the pressure of visit preparation on teachers.

Technology in the classroom

At the primary level as much as at any other level there is an important distinction between technology and science, which should be clear in the minds of curriculum planners and teachers, though it is probably not necessary for the children to be made aware of it at the primary stage. Science is concerned with understanding the world around us, whilst technology is concerned with using resources of materials, energy and natural phenomena to achieve some purpose relating to human activity. A central distinction is one of purpose; the main purpose of scientific activity is understanding, whilst solving a problem that meets a human need is the main purpose of technology, understanding not being necessary as long as a workable solution is found. There is, in practice, a close interaction of science and technology. For instance, in the course of scientific activity problems are often encountered which require technology, as in arranging a way to make particular observations (a microscope is a technical solution to such a problem) or in handling data (technology has produced the computer for this purpose), thus technology serves science in such cases. At the same time the solution of problems by technology involves the application of concepts arrived at through science (the understanding of reflection and refraction of light in the case of the microscope, for instance), thus science serves technology by providing concepts and principles to be applied.

It seems important to give children experience of both science and technology — of understanding and of problem solving — in the clear recognition of the different educational roles of these parts of the curriculum. The compromise of a mixture of the two is a poor development in the curriculum, for it gives neither the satisfaction of real problem solving nor the intellectual excitement of investigation for the purpose of understanding or satisfying curiosity. The compromise sometimes takes the form of science-oriented activities in which applications of science principles in everyday life are studied from the point of view of the illustration of the usefulness of science rather than the problem which was solved. In other instances it takes the form of a series of rather disconnected problems to be solved. Whilst each may be interesting, such activities do not on their own enable children to build up ideas which help their understanding and the solution of further problems.

As technology becomes more significant in the lives of more and more children it is necessary to consider its role in the curriculum carefully. For young children particularly (but for many older pupils as well) technology is best understood by participation. But this should not be at the expense of activities aimed at understanding. A balance is necessary. Table 1 attempts to describe the two kinds of activities — technological and scientific — which share an equally strong claim to be included in the curriculum. The description in each case is in the form of criteria to be met by activities suitable for primary children. They take into account the limitations in ways of thinking, in attention span and in experience of primary children.

TABLE 1

Types of activity relating to technology	Types of activity relating to science
• Problems which are understandable to children i.e. where they can see a clear need or benefit to themselves or others	• Investigations which give opportunity for basic concepts or ideas to be developed
• Problems which can be tackled in practice by children using simple equipment, tools and materials which are familiar to children	• Investigations which interest children and help to satisfy their curiosity
• Problems which involve children in using imagination and creativity in producing a unique solution (not in confirming a solution already known)	• Opportunities to interact with objects and events around them and so help the children gradually to make sense of their world
• Problems where there is the chance of a successful outcome in the time available and within a time span appropriate to the children's attention and concentration span	• Opportunities for children to use and develop investigative skills in testing out their ideas against evidence
	• Activities which foster scientific attitudes of respect for evidence, openmindedness and responsibility towards their living and non-living environment

Placed side by side in Table 1 the two lists tend to emphasise the differences rather than the similarities between the two types of activity. This may be helpful in maintaining a balanced view of curriculum content but perhaps unhelpful in creating unnecessary divisions. The ideas are offered for discussion and may well be disputed. However, if suitable technological activities are to be introduced in the primary school it is necessary to have guidelines which spell out their nature and what is meant by "suitable". Useful criteria must be capable of *excluding* some activities as well as providing grounds for *including* others. Thus, for example, a problem such as finding the optimum shape and number of blades in a windmill may be included as meeting the criteria proposed, whilst the problem of tuning an internal combustion engine would not meet three of the suggested points.

Conclusion

Many issues have been raised in this paper which urgently require further discussion and analysis. The following are among the questions raised which could usefully be addressed by all concerned with primary education.

1. Is there any support for the distinction between science and technology suggested here? If so in what agreed characteristics do they differ?
2. What educational values are there in activities relating to science and technology, such as those cited here or to be found in primary curriculum materials?
3. How can continuity and progression in the curriculum be accommodated in industry-based and technology-based activities?
4. What criteria should be applied in the selection of activities relating to technology and industry (see Table 1)?

References

1. Jamieson, Ian (Ed.) "*We make kettles*": *studying industry in the primary school.* Schools Council Programme 3, Longman, 1984.
2. Lewin, R. H., Technology alert — in the primary school. *School technology.* **15**, 2 issue 60, 1981

The following books provide a good sense of information on this topic.

The Unesco publication "New Trends in Primary School Education" Vol. I (ed. Harlen W.) covers many of the points raised in this paper. The publication is available in English, Arabic and French.
There is also a booklet published in Unesco's "Science and Technology Education Document Series" entitled *The Training of Primary School Educators* (ed. Harlen, W). It was published in 1985 and is based on the work of an ICSU-CTS sub-committee.

2

Industry and Technology in the Primary School: a Case Study from Indonesia

G. TANUPUTRA

IKIP, Bandung, Indonesia

Indonesia is an agricultural country. Eighty per cent of the Indonesian population live in rural areas, where most of them work in agricultural lands, cultivating crops such as rice, cassava, fruits, vegetables, tea and coffee. Many also work in industries varying in size from large ones employing hundreds of workers to small home industries. Agriculture is of direct relevance to all pupils and it is also a naturally integrating theme which has not as yet been represented in the typical science curriculum of Indonesian primary and secondary schools. What follows is a case study used in a primary school in West Java.

The study of a tea factory

This study is about a visit to a local tea factory by sixth-year pupils (aged 11–12) from a primary school in a village. The teacher concerned was implementing training received during an in-service workshop on the improvement of the primary curriculum.

The teacher planned the visit with the following objectives in mind:

(1) to familiarise the pupils with the real world outside the four walls of their classroom, in this case the industrial and technological world;
(2) to help pupils realise and appreciate the efforts and the energy put into the production of an agricultural product, in this case tea;
(3) to make the pupils understand work in a factory (for example, its organisation and structure, from the manager of the factory to the tea-pickers; the different tasks and different responsibilities of the people employed in the factory, etc);
(4) to give pupils an understanding of how tea is produced (selective picking of tea leaves, reselection of the picked leaves for the different

qualities of tea, processing, storing and packing for local use and for export);

(5) to link several subject areas through the study of industry and technology (science, social studies, mathematics, geography and language).

The teacher worked out an overall plan which consisted of pupils' and teacher's activities prior to the visit, during the visit and after the visit to the factory.

The first stage was a rough plan which included the objectives to be achieved by the pupils. The teacher discussed the ideas with his headmaster and colleagues to obtain suggestions and more ideas. Then he wrote a letter (approved and signed by the headmaster) to the manager of the factory. After obtaining agreement for the visit, the teacher made a preliminary visit to the factory. He was able to discuss and plan the arrangement of the visit in more detail with the person(s) in charge in the factory, go through a tour of the factory and talk to the people concerned with the visit, make a more precise plan which included points of interest and importance for the pupils, timing, and guidance for the pupils. After this visit the teacher revised the first rough plan.

Class activity before the visit

As an introduction to the study the teacher discussed with the children the different aspects of tea and tea production, as shown below.

Questions/statements/activities	*Relevance to content, skills and attitudes*
a. When and how do people drink tea? (occasions, sweetened or unsweetened, warm or cold). At this stage the teacher explained that tea is drunk in many parts of the world.	Daily life aspects, customs.
b. Do all countries in the world grow and produce their own tea?	Questions b, c and d bear relevance to geography, biology and commerce.
c. Which countries grow and produce tea?	
d. Why can tea be grown in certain countries and not in others?	

A world map is shown and the tea-growing countries are identified on the map (geographical position, climate, etc) which leads to discussion of the conditions for growing tea.

Skills: interpreting, pattern finding, comparing.

e. The teacher invites the pupils to study a tea shrub put in a pot (one pot for each group of pupils). Young leaves and old leaves are compared, the height of the plant is also measured. The plant is also classified (family, genus).

Biology.
Skills: observing, comparing, measuring

f. At what age is the tea plant ready for picking? The teacher did not intend that this question should be answered correctly, he wanted to make the pupils aware that there are other information sources which can be consulted. The pupils can get the information during the visit from the factory or from books or other sources.

Production process.

Attitude to evidence and sources of information

g. What is the average height of tea plants ready for picking? This will be investigated by the pupils during the visit.

Mathematics (statistics)

h. How is tea produced? The teacher outlined globally and briefly the production of tea. The details will be experienced directly by the pupils during the visit.

Technology

i. How is tea transported from the factory to the markets and abroad.

Transportation as a component in the link between industry and the world of commerce.

j. The interdependence of all the components involved in the production and marketing of tea was brought to the attention of the pupils during the discussion. A simple diagram was shown.

The world of industry and commerce; economy.

k. What are the tasks of women employees? What are the tasks of men? Most of the pickers are women, men do harder work such as loading and unloading trucks, operating the machinery and furnace, preparing the land for new tea plants — they get different wages. This will all be experienced directly by the pupils during the visit.

Social study.

l. How many kilogrammes of tea leaves are picked on the average by a tea picker per day? How much does a tea picker earn weekly? It depends on the amount of tea leaves picked during the week.

Social study and some mathematical aspects.

m. How many times a year is tea picked?

Questions m, n, o have relevance to production process.

n. When will young tea plants be planted in the field?

o. How do tea plants reproduce?

Biology.

p. Is the amount of tea produced each year the same? This question should be asked by the pupils during the visit. Production depends also on demand, crop failure. Mention diseases, research laboratories, pesticides.

Interdependence of components in the production process and marketing. Scientific aspects.

These questions/statements were put to the children as examples which might be of interest and importance to them and to serve as a broad guideline for the pupils as to what to observe, ask and note during their visit. Other questions might come up from the pupils themselves. Their curiosity should not be inhibited.

Pupils were divided into small groups of four to five, each with a different focus of attention. Each group was asked to prepare questions and notes of what to observe during the visit. The pupils were also briefed about orderly behaviour and some safety measures were explained. They were asked to take with them on the visit necessary things such as notebook, pencil, hat, packed lunches and drinks.

Activities during the visits

All aspects and questions during the discussions were asked and responses noted by each individual group, especially those related to their own task. It did not mean, however, that other groups were not allowed to ask certain questions. The visit started from the hills outside the factory where the pupils examined the tea plants in more detail, measuring a certain number of tea plants of the same age. They also saw how the women picked particular tea leaves and asked the women questions such as how much tea they pick in one day, how many hours they work in a day, and so on.

The groups of pupils then moved together to the factory. Although each group had a different task, it was felt important that all pupils went through the whole activity during the visit.

Activities after the visit

The follow-up work began another day, because the visit itself took up a whole day. To allow time for the pupils to sort out their notes and prepare their report, this activity was carried out a few days after the visit. It consisted of the following.

Each group reported to the whole class. Whilst one group reported, the other groups listened and made notes. The other groups then contributed, making suggestions or comments on the report. The role of the teacher in this sesssion was as "manager". He sometimes intervened to clarify doubts or uncertainties or sometimes probed to get out more from what the pupils have observed. He also made links between the different subjects involved (see Fig. 1).

Each group then made a display of its own finished task. For example, one group made a diorama of the hills where the plants are grown. Another group drew a flowchart of the tea-production process. Other display materials made by the pupils were:

— a world map with the tea-growing countries indicated and also the countries to which tea is exported,
— a drawing of the scene of women, with their large round straw hats, picking tea leaves,

— a drawing of some of the machinery used in the factory,
— a table of statistics of the production of tea over a certain period, including a graph derived from it,
— a display of lots of tea plants of different ages,
— a schematic diagram of the linkage between the different components involved in the production and marketing of tea.

The displays were mounted on the walls in the classroom where they were kept for some time. Pupils from other classes were given the opportunity to see the display.

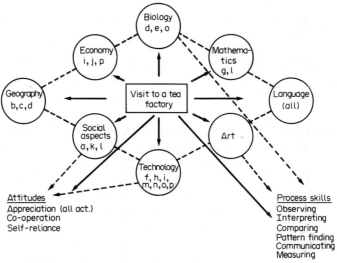

FIG. 1

3
Using Elementary Technology to Teach Primary Science

P. VIOLINO

Universita di Pisa, Italy

The introduction of a specific technology into the primary school classroom can be used as that concrete experience which leads children to an understanding of an important scientific concept, such as first ideas about energy.[1]

The subject of the Industrial Revolution is taught in Italy at the beginning of the fifth class, and it is natural to propose to the children an investigation about engines (particularly the steam engine) and their technology. The teacher suggests, however, that it is convenient to begin with a simpler device, the overshot water wheel.[2] After some preliminary description by the teacher and some suggestions, a first very simple design is produced by the children themselves. The teacher suggests what to do, shows how to use tools and materials and sometimes gives practical advice, but allows the children to make their own design mistakes, only to find out later what went wrong. The children learn some very simple techniques, that may be new for them but not strange, such as cutting plywood and iron sheet, making holes, driving nails, and so on. The teacher insists on accuracy but not on aesthetic quality. After the first wheel is assembled, it very seldom works properly but this is not a problem since the children very quickly realise the cause, that may be either in the design, or in the building stage, or both. After a number of successive improvements, they obtain a properly operating wheel, that does some useful work like raising something from the floor.

However, a local waterfall is not always available for testing the design. An alternative approach is to consider the steam turbine. The same wheel can still be used, and as an obvious steam source a coffee pot for making Italian Espresso. This is not dangerous since the pot has a built-in safety valve and its technology is quite straightforward. The turbine rotates very well, but it is quite clear that most steam gets lost — solution — send it into a closed chamber. The children are not able enough to build a system with cylinder and piston, but again a quite simple, low-cost and well-known device can be found; a pump for inflating bicycle tyres. It has only to be modified by replacing the leather gasket with a metal, lubricated gasket. By sending the steam in it, the piston is raised, and the system works roughly as a Newcomen engine, but without a condenser. Thereafter a working model of a Newcomen engine is presented to the children who, notwithstanding the enormous difference in appearance from their coffee-pot and bicycle-pump machine, instantly recognise which parts do the same job in "their" engine and in the "real" engine.

The enthusiasm of the children leads them to build a number of further machines; windmills and/or an electric motor made out of scrap wood, nails and some copper wire, but working properly.

Clearly, between such activities, a considerable time is given to discussions about the physical principles underlying the experimental work and also to further activities (for example, about magnetic and electromagnetic interactions) before designing the electric motor.

It is after such activities that the children are led to consider the similarities between what they are doing and many common-place devices. They also find out that in order to use a device we need "something else" to put into it. Eventually they realise that all these devices perform transformations:

height of falling water	INTO	movement
heat	INTO	movement
electricity	INTO	movement
electricity	INTO	heat
movement	INTO	electricity
movement	INTO	heat

When this happens, it seems natural to find a common name for the "agent" of the transformations and the teacher suggests "energy" and from then on the children use the word "energy" in an essentially correct way without having ever heard a formal definition.

Conclusion

The above is just one example. Simple technology has also been used by the children to build model houses and furniture for them where a correctly proportioned reproduction is essential for all parts to fit together: to build simple instruments (a sensitive balance, thermometer, water clock, rain gauge, . . .); to sew coloured thread patterns whose symmetry properties are being investigated; and so on[3a–f]. Many more ideas can be developed along these lines.

The children enjoy this kind of activity, and they really appreciate the underlying scientific ideas, since they not only operate at the concrete level, but they are also involved intellectually in all stages of the work. They are thus led to think that science is fun, and that, I believe, is a considerable result.

References

1. Violino, P., *La Fisica nella Scuola*, **11**, 6, 1978.
2. VITA, Development Digest, October 1975, p. 55.
3. (a) Bosman, L., Lazzeri, F., Legitimo, J., Martinelli, L., and Violino, P., *Le Scienze, la Matematica e il loro insegnamento*, **18**, 287, 1981;
   (b) Bosman, L., Lazzeri, F., Legitimo, J., Martinelli, L., and Violino, P., *La Fisica nella Scuola*, **15**, 80, 1982;
   (c) Violino, P., *European Journal of Science Education*, **4**, 115, 1982;
   (d) Bosman, L., and Legitimo, J., *L'Insegnamento della Matematica e delle Scienze integrate*, **4**, 7, 1983;
   (e) Bosman, L., *L'Insegnamento della Matematica e delle Scienze integrate*, **6**, 28, 1983;
   (f) Bosman, L., and Giachette, L., *L'Insegnamento della Matematica e delle Scienze integrate*, **7**, 11, 1984.

Section B

Industrial and Technological Issues in Secondary Science Curricula: Setting the Scene

Introduction

No matter in which region of the world we live, the products of science and technology are increasingly entering our homes, our schools and our workplaces. As a result, there is a growing concern to incorporate technological and industrial examples into school education as naturally and effectively as possible. This section is concerned particularly with secondary science education. At this stage, students usually have enough background in mathematics and science to place industrial and technological themes in context. Furthermore they will shortly be involved in important career choices. The reasons for including such themes in school (and tertiary level) curricula are numerous and include our desire to

— increase the scientific and technological literacy of the public at large, and in particular those who have responsibility for decision making, so they may better exercise their daily, community and local or national political and economic responsibilities,

— give individuals practical understanding as well as knowledge of the new technologies which impinge on their lives,

— give a good preparation for those who are choosing an industrial and technological career,

— motivate students as relevance to the real world is essential to student interest and learning,

— educate future decision makers.

In this section, Steward and Towse describe some of the difficulties in incorporating industrial and technological themes into curricula and point to interesting developments which they instituted. Holman then outlines some general principles, with examples, on the strategies that can be adopted in this area of curriculum development and Lazonby sounds a cautionary note: we must not simply assume that the inclusion of these topics are, *per se*, interesting to students.

4

Incorporation of Industry and Technology into the Teaching of Science

J. W. STEWARD
University of Papua New Guinea

P. J. TOWSE
University of Zimbabwe

Why should industry and technology influence science teaching?

Much science teaching is restricted to a rather rigid diet of facts and concepts, conveyed by teachers relying largely on the use of "chalk and talk". For many students, science is seen as relevant only within the confines of the school laboratory (if there is one) and bearing no relationship to their everyday lives. This view can only help to perpetuate the idea that the teaching and learning of science are concerned solely with the facts and concepts defined by syllabuses and examinations. It should be recognised that many students who pursue science courses in secondary schools will not go on to further academic studies in the sciences and that a narrow academic view of science is not appropriate for such students. We would also question whether it is appropriate for those who will pursue science subjects at a more advanced level. As the impact of science and technology on their lives spreads at an ever-increasing rate, it is important for as many citizens as possible to be exposed to some of the wider issues arising from that impact.

This does not mean that everyone should be an expert on science and technology — but at least everyone should have the opportunity to think about the issues. Even if one does not know the answers, one ought to be in a position to ask at least some of the right questions. To hold well-informed and socially-responsible views one needs to understand at least some of the issues involved in the use and abuse of one's natural resources. Neither the Minister of Finance nor the villager needs to know the technical details of paper making, for example. But both need to be able to question the possible impact on the environment of a large paper-making plant.

There will be many problems to face, given the knowledge and skills to be acquired by teachers and the lack of scientifically based industries in many Third World countries — but there is need to make a start.

What strategy is proposed?

Syllabuses used in many parts of the world have usually included industrial applications as a very small part of the total content, particularly in chemistry. The Caribbean Examinations Council has recently published a new chemistry syllabus which contains a separate section devoted to chemistry and industry. Kenyan syllabuses have extensive references to the applications of chemistry. But even if syllabuses are changing, teachers still need to be convinced that the changes are for the better and to be encouraged to teach the new ideas. If the changes merely mean a little more teaching about a few industrial processes, then they will have been in vain. There need to be changes in attitude, skills and knowledge. For effective teaching and learning about science and technology in secondary schools, teachers need to be exposed to ideas about the applications of science and technology.

Thus we propose that the pre-service training of all teachers, in universities or in teacher training colleges, should include a course in industrial applications and some practical experience of industrial processes. The implications for teacher training are significant, but it should not be too difficult to incorporate these components into the pre-service training of secondary school science teachers. The course in a particular country should look broadly at the industries in that country and concern itself as much with the social, economic and political issues of these industries as with the scientific and technological ones. It should examine such issues within a framework of national, regional, continental and universal activity.

The value of industrial experience cannot be overestimated[1]. If it is not possible to have extensive first-hand experience of a variety of industries, then a good second best would be the undertaking of at least one in-depth case study of local industry. It should be possible to arrange for students to be attached to particular industries for a limited period. They would have to be very clear about the basic scientific principles involved, but they would need to be able to see these within a broader framework of other issues. For example, the merits of one method of manufacture as against another, the merits of batch processes as against continuous processes, the potential impact of the different processes on the environment, the economics of the supply of raw materials and so on. For the maximum use to be made of the resources available in a particular country, there needs to be a planned programme of links between education and industry so that students obtain as broad a view of industry as possible. Students offering chemistry as a major subject at Mico College, a teacher's college in Jamaica, have to submit an industrial case study. BEd students at the University of Zimbabwe take a course in industrial chemistry and have some experience of industry.

The much greater problem is that of in-service training for established teachers and for graduates who enter teaching without taking a post-graduate certificate course in education. The ideal solution would be for them to be released from schools to take part in the same course and in the same industrial experience scheme as the pre-service students. If their release during term time were not possible, there are always the vacations and, providing the incentives are strong enough, say, promotion to a higher point on a salary scale, we believe that teachers would be willing to participate.

The logical extrapolation of this proposal is that those teaching more advanced courses would need greater exposure to applications. The nature of such exposure depends on the syllabuses to be followed, but more importantly on the number of industries available locally. It should be seen as an important part of both undergraduate courses and teacher training courses.

It should be clear that we favour the inclusion of a study of appropriate industries in all university science courses. This means the closest possible co-operation with the industries concerned, but experience in Zimbabwe suggests that those in industry are only too willing to offer teachers the opportunity to clear away the "general fog of ignorance" in schools as to what goes on in industry. In fact, one of the more radical implications of our proposals would be the much greater involvement of those in industry with syllabus construction at all levels.

However, we do not see all this simply as extending the "uses and applications" aspects of existing syllabuses. There are so many issues involved in the impact of science and technology on the socio-economic welfare of a country that there is a need to make that impact a more central concern of science teaching.

What resources are needed?

Before changes can be implemented, we need to consider what resources are needed. It is not enough to provide basic school texts, for the teacher has to be more than a mere dispenser of information. Resources are needed for teacher training and for teaching within the secondary school, and it will take time to build up such resources. We suggest these resources include: (i) information booklets, (ii) tape/slide sequences, (iii) simulations (iv) computer-assisted learning.

For effective development of some of these resources, collaboration is needed between education and industry, those in industry providing at least the basic material for teacher trainers and teachers to edit. There is also the vexed question of the cost of producing such resources.

Resources produced by industry in industrialised countries have in the past been widely available, including some excellent free materials from organisations such as BP, ICI, Shell and Unilever[2,3]. Resources for teachers in developing countries are much scarcer. For example, there have been occasional articles[4,5,6,7] but some of these are now a little dated. Very few school texts include references to specific examples of industry.

Conscious of just how little material there really is in Africa, we have written a resource book for chemistry teachers[8] which, as far as we know, is the first to look at chemical industry on a continental basis. The irony is that it is often easier to obtain accurate information about African chemical industries in London than it is in Africa itself. Our hope is that teachers may be encouraged to develop their own resource materials, using some of the ideas in our book as a starting point. A beginning would be the compilation of files of newspaper and magazine cuttings relevant to industrial issues of national importance.

Experience has shown that games and simulations provide great motivation and encourage a high degree of participation. *Buena-fortuna* and other simulations developed for the *Science in Society* project[9] have few counterparts in the Third World, although some developed in Zimbabwe will be described later, on page 334. Some computer simulations developed in the industrialised countries can be used universally, but there is little designed specifically for the Third World. Computer-assisted learning has had mixed success, largely because of a lack of good materials. Nevertheless, with hardware costs falling, microcomputers are beginning to find their way into classrooms in the Third World and the problem is essentially one of developing appropriate software.

References

1. Morgan, D. R., A teacher in industry: a Morall experience, *Educ. Chem*, 1979, **16**, 15–17.
2. Ball, C. J., Frazer, M. J. and Haines, B. A. Aids to school from the chemical industry, *Educ. Chem.*, 1975, **11**, 150–151.
3. Childs, P. E., Chemical literature for the teacher, *Educ. Chem.*, 1979, **10**, 49–51.
4. Allsop, R. A., Freeman, J. M. and Ingle, R. B., Chemical industry in developing countries, *Educ. Chem.*, 1971, **8**, 226–228.
5. Haden, J., Iron and education in Uganda, *Educ. Chem.*, 1973, **10**, 49–51.
6. Huxley, J. V., Recent developments in copper production, *Educ. Chem.*, 1973, **10**, 94–97.
7. Morgan, D. R., Salt production in Tanzania, *Educ. Chem.*, 1971, **8**, 20–21.
8. Steward, J. W. and Towse, P. J. *Chemical technology in Africa*. Cambridge: Cambridge University Press, 1984.
9. ASE, *Science in Society*. London: Heinemann Educational Books, 1981–1984.

5

Contrasting Approaches to the Introduction of Industry and Technology into the Secondary Science Curriculum

J. S. HOLMAN
Watford Grammar School, Watford WD1 7JF, UK

Although there is a demand to make secondary science curricula more relevant, the justification for relevance varies. If one views the importance of science education in terms of training a workforce, then industrial and technological issues are important. However, there are many who see science education as an integral part of a general education where the importance of relevant science is to help students to understand the place of science in the world about them. Such a perspective is likely to stress the social issues of science as well as its implications. Finally, there is the need to interest and motivate students and many teachers feel that science taught within the framework of relevance may do this more successfully than courses which are designed around abstract concepts.

If the science teachers are to make their courses more relevant to society, then a pressing need is for teaching resources. There are two contrasting approaches to the development of such courses, which the author discusses in this paper.

Science first, or applications first?

The first approach starts with existing science courses, pure and lacking relevance as they often are, and builds in relevant applications or issues at appropriate points.

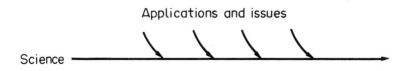

One might describe this evolutionary, "Science-first" approach as "dropping relevance into the interstices" of existing science courses. It is likely to require back-up resources, probably of a modular nature, but not completely new courses.

The second approach is more revolutionary. It starts with the applications or issues and develops the relevant science from them.

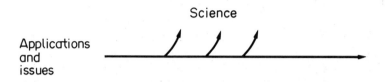

For example, one might begin with the topic of Clothing and from it develop scientific ideas such as heat transfer and polymers. One is "digging out" the science that lies within all everyday topics. This second, "Applications-first" approach implies the abandonment of existing courses and the creation of wholly new ones.

"Science-first" exemplified: the Science and Technology in Society project

The English Science and Technology in Society project (SATIS) illustrates the "Science-first" approach. It is a new initiative of the Association for Science Education in which a team of teachers developed resource units. These units are short, occupying no more than 75 minutes of classroom time, and designed for quick, off-the-shelf use. Each has a clearly recognisable link to a topic commonly covered in science courses for 13 to 16 year-old children.

The units all involve some form of interaction. The project recognises that written material alone, however readable, well-presented, and well-illustrated, is unlikely to involve pupils at more than a superficial level. The units therefore aim to involve pupils in some kind of interactive activity — whether it be simply answering questions, or more complex structured activities such as role-playing, data-handling, decision-making, class discussion and so on.

For example, the unit on Nuclear Power is intended to be used in conjunction with work on nuclear structure in physics, chemistry or integrated science lessons. The unit comprises a structured discussion, and begins with a simple general briefing on the principles of nuclear fission, which children might study for homework on a preceding evening. Pupils are then divided into small groups of five or so, each with a chairperson. Each pupil in the group has an "Expert's Briefing" on some aspect of nuclear power, and in the ensuing group discussion each child contributes information and opinions on the nuclear power question. In this structured way, each pupil must contribute and become involved with the activity.

The Industrial Gases unit deals with some of the important gases commonly met in a chemistry course, but instead of looking at their preparation and properties, they are examined from the point of view of their industrial manufacture and use. Pupils are given figures on cost, volume of production and uses of the gases, and asked questions which require them to handle and evaluate this information.

The unit entitled "Microbes make human insulin" introduces some simple ideas relating to biotechnology, and might be used in conjunction with work on genes, chromosomes and DNA. Information is given on the technique of gene manipulation and gene splicing, and the application of these techniques to the bacterial production of human insulin is described. Questions follow which explore the ideas and look at some of the technological and ethical problems involved in "genetic engineering".

These three examples should serve to illustrate the kind of resource units being produced by SATIS. Further examples are listed below. There will eventually be up to a hundred units, giving a wide coverage of the science curriculum and dealing with a wide range of applications of science and technology.

Title	Description
Sulphurcrete	Information, discussion and practical work on the use of sulphur as a building material
A medicine to control bilharzia	Information, questions and discussion on a case study of the development of a pharmaceutical product
What's in our food?	Homework survey and subsequent analysis relating to the use of food additives
Cross-Channel link	Decision-making exercise concerning the construction of a bridge or tunnel across the English Channel to France
Fibre optics and telecommunications	Background information on the principles and uses of optical fibres
Dam Problems	Role-playing simulation concerning environmental problems involved in building a hydroelectric scheme
The Bigger the Better?	Data analysis, decision-making and discussion concerning the optimum size for a petrochemical cracker and other industrial units
The Design Game	A game based on the design of an energy-efficient house

Assessment is considered to be a key area for action by the SATIS project, for two reasons. First, this is a relatively new area for most science teachers and examiners, accustomed as they are to working in the apparently neutral and value-free realm of pure science. There is little common experience of assessing pupils' knowledge and awareness of the applications and issues of science, and work needs to be done on the development of assessment techniques. Second, it would be naive to suppose that examinations do not have a profound effect on what is taught in English schools. It is thus important to demonstrate that the means exist to assess this area of science education. Some examples of assessment items related to SATIS units are given in Chapter 37.

In addition to the pupils' resources, a General Guide for Teachers' is available. This contains background information and general guidance for the teacher. For example, it includes sections on visits, suggesting how industrial visits can be made most effective, on using the local environment for "science trails", on the use of newspapers and magazines as the basis for science lessons, and so on. There is guidance on how to organise class discussions, suggestions for problem-solving activities, and useful resources such as books and films will be identified. Both General Guide for Teachers' and Pupils' resources were published in Summer, 1986.

Thus the Science and Technology in Society project takes a pragmatic approach, starting with science courses as they now are, and adding social, industrial and technological enrichment. The Salters' Chemistry course is an example of the more revolutionary "Applications-first" approach.

"Applications-first" expemplified: the Salters' Chemistry course

Concern over the popularity of chemistry in schools, and its future as a separate subject in the face of trends towards integrated science, gave the stimulus for this radical reappraisal of school chemistry. Working at York University, a team of teachers and others involved in chemical education are creating a chemistry course from the bottom up. Rather than adapting or modifying existing chemistry courses, the decision was made to start afresh, and to begin not with chemical facts and principles, but with everyday experiences familiar to school children.

Work began in September 1983 on course material for children aged 13–14. Intensive activity led to the production of five modules, each based on an important area of everyday experience: Food, Drinks, Warmth, Metals, and Clothing. Each module is designed to occupy between five and seven 75-minute periods.

True to the philosophy of "Applications-first", the modules were created by reference to everyday themes, not to preconceptions about what should be taught in a chemistry syllabus. Yet they demonstrate clearly the ease with which scientific principles can be developed from everyday applications. Thus, for example, in the Drinks unit, carbon dioxide makes an appearance, as it does in all chemistry courses, but in this case it is first met in the familiar context of fizzy drinks instead of the usual laboratory gas generation experiments.

Work continued in 1984 on the course material for 14–15 year-olds. Once again, the approach was modular, and this time seven modules were developed: Agriculture, Transport, Buildings, Emulsions, Minerals, Plastics and Food Processing. As before, the "Applications-first" principle was followed, with no reference to preconceived ideas as to what a chemistry course should contain, other than the work already developed for the previous year of the course. Important chemical themes again emerged: for example, in the Buildings unit the themes of reaction rates and the relation of properties to structure are developed from a consideration of common building materials such as stone, metal, wood and brick.

With the development of modules for 15–16 year-olds in 1985, the course was completed (in English schools, compulsory education finishes when pupils are aged 16). There has followed a long period of school trials, revision and editing, and it is intended to publish the course by 1988[2].

Thus the Salters' Chemistry course takes a radical approach, starting with the everyday interests and experiences of pupils, and developing chemical themes from these. How does this radical approach compare with the more pragmatic approach exemplified by Science and Technology in Society? What are the strengths and weaknesses of each?

The two approaches compared

Imagine a school science department, anxious to make their courses more relevant to industry, technology and the world outside the classroom. Which approach should they adopt? A number of questions are likely to be asked as they try to reach their decision.

1. *Which is easier to implement?* There is little doubt that the evolutionary "Science-first" method will be easier to put into practice, since it seeks to modify existing courses, rather than creating new ones.

2. *Which will be more acceptable to pupils?* This will naturally depend on the pupils. However, the "Applications-first" approach, beginning with the interests and experiences of pupils, is likely to be more successful at drawing their attention, at least initially.

3. *Which will be more acceptable to teachers?* Relevant teaching tends to involve the teacher in a change of role. From the traditional role of the science teacher as a purveyor of apparently neutral and value-free facts and concepts, the teacher is likely to move into an area of value-judgements, where there is not always a single, right answer. By inviting pupils to consider the question of nuclear power, for example, the teacher is inviting them to offer their opinions on the nuclear question, as well as considering

the facts of nuclear fission. The teacher may become more of a neutral chairman than a dispenser of facts, and this change of role can be threatening. Teachers are likely to need help and guidance. They will also, of course, need resources, because few science teachers have the depth of knowledge and experience of industry and technology that is needed in order to place the subject in context.

These problems will arise whichever approach is adopted, but it is likely that some teachers will find the more gradual "Science-first" approach less threatening — though others, impatient for change, may feel it to be too cautious.

4. *Which will be a more effective way of placing science in its wider context?*   Teachers often feel under time pressure, particularly when they are following examination courses. Under this pressure, the tendency is to concentrate on what is felt to be essential, and in this situation it is likely to be the applications that teachers omit. The "Applications-first" approach, with applications fully integrated into the course, is more able to resist such backsliding. Furthermore, this approach embraces wider educational objectives in terms of skills developed and contexts encountered.

5. *Which will be more effective in developing scientific knowledge and understanding?*   Most science teachers regard their major task as being in the cognitive domain, and for them the coherent development of scientific ideas may be of the greatest importance. The "Applications-first" approach is always open to the challenge *"But is it science?"*, and certainly it is important to strike the right balance between applications, issues and scientific ideas. There is also the problem of coherence: with "Applications-first", scientific ideas are likely to arise in a fairly random and haphazard way, and probably not in the order that the teacher would regard as a logical development. This makes it important to design modules carefully, and to think carefully about the order in which they are followed.

However, the teacher's idea of a logical development of the subject may not necessarily make the same sense to pupils, and for children it may be more important that the ideas appear in a familiar context than in a particular order. Furthermore, a modular course based on the "Applications-first" approach allows a particular idea to be developed in one module, then revisited later in another module. This spiral development may be more effective at establishing and reinforcing ideas than the traditional, linear approach.

Realistic teachers accept that teaching is an inherently inefficient process. Children rarely give the lesson their full attention, and often they hardly give any. The motivation provided by relevant science helps, but but even so most children are likely to miss parts of any topic, whether through inattention, absence or other causes. Many, perhaps most, children fall by the wayside as the development of the topic becomes more difficult, and many will never reach the summit of its culmination. Given this, it is perhaps better that they fall by the wayside in the possession of some knowledge and ideas that will be relevant to their future lives, rather than in the possession of a few abstract scientific facts and ideas. In other words, for the pupil who only succeeds in travelling part of the way, the "Applications-first" approach may be more relevant than "Science-first".

References

1. SATIS materials are obtainable from the Association for Science Education, College Lane, Hatfield, Hertfordshire AL10 9AA, UK.
2. Salters' Chemistry materials are available from the Department of Chemistry, University of York, York YO1 5DD, UK.

6

Do Students Want to Learn About Industry?

J. N. LAZONBY
University of York, York, UK

It is essential that we are able to justify the place of science within the total curriculum, not simply assume that it should be taught. There are constraining pressures on the curriculum in secondary schools. On one side, there are new subjects such as Computer Science and Electronics. On the other, there are creative subjects such as Art, Music and Crafts which demand inclusion.

In search of the justification, we can consider what applied aspects of science have to offer. These have a bearing on all the areas studied at Bangalore, not least those concerned with industry. Industrial applications can be considered in terms of the scientific principles of the process, the technology of the process and the broader aspects, those factors influencing the decisions to use it and the implications for society.

However, an assumption which often is implied is that students will find the industrial aspects more *interesting* than "pure" science. The work of several teachers provides a warning about this assumption. Coombes, in a study on some 15–16 year-old students in the UK[1], found that both teachers and students supported the idea of including more industrial applications and their social implications in their curriculum. However, they did not find these aspects more interesting. Metcalfe[2] in another study, this time on 17–18 year-old students, found a similar reaction. He also found that chemistry students who hoped to enter the chemical industry on completion of their education, were less interested in the broader social and economic aspects of industrial processes than those who had no intention of entering the chemical industry.

Further work done by Kempa and Dube[3] in the UK and by Chan[4] in Singapore showed that chemistry students (16–18 years-old) who were high achievers preferred studying the principles rather than the applications of the subject. Indeed, Carter[5] in the UK showed that this difference was even more marked where the applications were clearly concerned with "applications to the outside world", rather than "academic applications".

Thus, we should not argue for the inclusion of industrial applications on the grounds that they make the subject more interesting. We must justify their inclusion in general educational terms. Further, the challenge, particularly for those developing curricula for the high achievers, is to make these courses are at least as interesting for them as the traditional "pure" science course.

References

1. Coombes, S. D., Lazonby, J. N. and Waddington, D. J. *Education in Chemistry*, **17**, 6 (1982).
2. Metcalf, I. M.Sc thesis, University of York , 1980.
3. Kempa, R. F. and Dube, G. E., *British Journal of Educational Psychology*, **43**, 279 (1973).
4. Chan, O. L. M. M.Sc thesis, University of York, 1984.
5. Carter, G. E. *British Journal of Psychology*, **52**, 378 (1982).

7

Discussion

It was generally agreed that in principle the Applications-first approach described on page 32 was the most promising, but the most difficult to implement. Variations on these approaches were also identified. One in which applications are hooked-on is the explicit reference by the science teacher to appropriate industrial applications when the topics and concepts of science are taught. The assistance which teachers need is examples of local industrial applications related to topics in the science curriculum with sufficient detail so that teachers can make these references interesting, clear and socially significant. Newspapers, TV and other media are often helpful in this respect, but teachers need guidance in their effective use.

Another approach discussed was the hooking-on of technology and social aspects by taking students on short learning expeditions associated with regular topics in the science curriculum. These learning excursions should be based on simple but accurate details of technological applications including their social implications. They also enable a number of science topics in the curriculum to be related together and reinforced. The attractions of this approach are the increased motivation of students and the more successful learning which stems from the wider context. The teachers need indications of those points in the curriculum where such activities are appropriate.

A further approach discussed was adding social content through examination questions. This accepts that in many countries examinations are a powerful influence on the curriculum. It calls for questions which assess knowledge and skills presented in industrial or other social contexts, and examples are needed of such test items for teachers to use as models. Guidance on the development of such questions is needed, along with appropriate in-service training.

The last approach discussed was that in which knowledge and skills are used to solve problems, including industrial ones. Such an approach helps teachers and students to recognise that technologies exist to solve real problems.

Science knowledge
Science skills
→ Technology ←→ Solving a real life problem

This sequence can go either way as teachers gain confidence. The help which teachers need is suggestions of real problems which could be used, along with appropriate teaching materials. Once again, in-service training is important.

All the approaches above are variations of the "Science-first" approach. If "Applications-first" is likely to be more effective in teaching about the science and technology which will dominate the lives of our students, will we have long to wait?

Technology in the Secondary Science Curriculum

Introduction

In many countries, science and technology education have tended to develop along separate lines. This leads to problems of duplication and hence overcrowding of the school curriculum. Choices have to be made between "science" and "technology" as the subject to be studied. In a recent paper, *In place of confusion*, Black and Harrison[1] attempt to resolve this problem by adopting a model for the curriculum as a whole within which the distinctive roles of science and technology are indicated.

In the papers in this section, the most insistent argument is that we should appreciate that the teaching of science and technology are part of a continuum. This is particularly apparent in the papers by McKim and Swift.

Allsop sets the scene for us, looking at different strategies of including technology in the curriculum. Swift discusses a physics course based on rural technology and Mehta describes how relevant science and technology can be integrated. Ibrahim describes some activities for the rural poor, the most disadvantaged segment of the Bangladesh society, emphasising that the rural technology he discusses may sound commonplace but is a matter of life to the rural multitude.

At the other extreme, Kille and Johnsen look at ways in which biotechnology can be included in the curriculum. The recent publicity given to genetically-based technology — what is glibly termed genetic engineering — hides the fact that the science of biology has a long-standing relationship with technology. Agriculture, medicine, pharmacy and many other areas of industry have literally for centuries benefited from, and been a spur to, biological sciences. This broad view of biotechnology is portrayed in these two contributions. It is important that this should be so because, for developing countries in particular, one needs to recognise that many of the needs of development — health, food, energy, for example — depend greatly on biological resources, and hence on biotechnologies, if they are to be met. Johnsen provides a review of the nature of biotechnology, concentrating on the school curriculum. Kille reviews the education and training needs for biotechnology at all levels of education. He stresses both the need to maintain courses in basic biological topics and to introduce ones concerned with the processes of industry such as production and marketing, particularly in relation to the newer bio-technologies.

Holbrook, like Mehta, gives specific examples of the ways in which science and technical skills can be involved in a single activity. Nwana looks at the possibility of introducing technology into the science curriculum in his country, Nigeria, and the constraints he faces.

McKim's paper is concerned with an approach of infusing technology into a conventional academic science curriculum, describing the Physics

Plus project. Sun describes some work at her school in China, which takes the form of extra-curricular activities in which students are encouraged to study technological problems and their solutions. Chandavimol looks at science and technology in education through an entirely different viewpoint — and we have a description of the inclusion of science in a vocational education programme in Thailand. Mottier and Raat indicate ways in which another minority in science and technology education — girls and women — can be helped, giving examples from the Netherlands.

Reference

1. Black, P. and Harrison, G. *In Place of Confusion: Technology and Science in the School Curriculum.* Nuffield-Chelsea Curriculum Trust and the National Centre for School Technology, Trent Polytechnic. This paper is available from the Association for Science Education, College Lane, Hatfield, AL10 9AA, UK. The price is £2.

8

Factors Affecting the Uptake of Technology in Schools

R. T. ALLSOP
University of Oxford, UK

This contribution is concerned with a research project, based in the Department of Educational Studies in Oxford University, between 1980 and 1984 in which the introduction of technology was studied for the two age ranges 11–13 and 14–16 in Oxfordshire schools.

In the first 2 years of secondary schooling in England, pupils (aged 11–13) normally take a common curriculum which includes science and CDT. In CDT (craft-design-technology) the emphasis in on craft-design, the use of a variety of materials and introducing graphic skills. It would be unusual for a systematic technological flavour to be added for this age group so the research team worked with groups of local teachers to produce curriculum materials relating to all aspects of technology but particularly initially concentrating on problem-solving projects. We introduced the mnemonic PRIME (Problem – Research – Ideas – Make – Evaluate) in order to assist younger pupils with understanding the technological process. We studied the development, reception and effect of these new inputs into the science and CDT curricula in the schools. Major findings for the 11–13 age range are summarised below.

Teacher/curriculum factors

Where teachers perceived science as a discrete body of knowledge, the school science curriculum as similarly constituted, and were under pressure from examination requirements higher up the school, then they were unlikely to introduce problem-solving technological activities. Science teachers who perceived their teaching as being concerned with *processes* more readily incorporated technology in their teaching.

We found a variety of ways in which teachers were able to introduce problem-solving work, either by infusion into existing topics or as self-contained activities strategically placed throughout the course. Teachers of CDT were much less likely to be hampered by subject structure or by examination requirements and thus were likely to be more open in trying projects. However, at this level, they were more likely to insist that an essential part of their work was to instil basic skills in the working of materials (traditionally wood and metal) *prior* to their incorporation in problem-solving project work. CDT teachers were also less likely to repeat well-tried project work with which they were bored — "Once we've found the optimum solution to a problem, it is time to try something new".

Resource factors

There are currently few resources for teachers wishing to introduce problem-solving work to this age range. Collaborative activity in developing ideas seems likely to be much more fruitful that the production of formal, written materials. Many of the ideas for problem-solving seem to come from the area of physics — the project found it difficult to generate viable technological projects with a biological or chemical emphasis.

It was always envisaged that such problem-solving projects would use simple apparatus, rather than sophisticated laboratory equipment. There is undoubtedly some resistance to this, on the part of both teachers and pupils, once science laboratory work is linked to expectations of using relatively sophisticated equipment (this seems equally true once conventional laboratories have been equipped in Third World countries). Perhaps Rutherford was right when he said, "We haven't the money, so we shall have to think!". Nevertheless, the provision of resources sufficient for 250 pupils aged 11 in one year group to carry out problem-solving projects remains a serious consideration!

To investigate technology in the 14-16 age range, case studies were undertaken at five schools where Technology was introduced as a subject leading to a specific public examination at age 16. The initiation of such a programme has been made much easier by the recent publication of the Schools' Council Modular Technology course[1]. We studied schools where the course in technology was being organised by science teachers, by CDT teachers, and by science/CDT teachers working as a team. The major findings of this part of the research, and some of the questions which were raised, are summarised below:

(a) Those who started teaching a technology course were enthusiasts whose personal interests and commitment enabled them to surmount the difficulties of introducing a new subject into the curriculum. (Can technology courses survive if they depend on enthusiasts who may move on, leaving a school with no one interested in continuing the course?).

(b) One of the major facilitating factors in starting the technology courses in our schools was the support of their headteachers, who appeared to be responding to current calls for reform of the curriculum for 14–16 year-olds, and for higher quality recruits for technologically-oriented courses in higher education.

(c) The technology courses were offered to students of higher ability and linked to examinations at age 16. This immediately secured for them a recognised place in the curriculum. (Is a separate technology course the best way to introduce technology into an already overcrowded curriculum? Will marketing technology towards higher ability pupils ensure status, and thereby survival, for the subject? Will it alter the traditional image of craft/technical subjects as being for the lower ability pupil? What technology provision should be made for the less able?).

(d) Resources for teaching technology in this age range are expensive and difficult to obtain during a time of economic stringency. Some of the difficulties which this might create have been overcome by (a) and (b) above.

(e) Technology was justified as a subject for inclusion in the curriculum on the grounds that "it allows the opportunity to solve problems" and "it fosters thinking, independence and self-organisation and a critical appreciation of the importance and impact of technology in everyday life". (At age 14–16, is it too late to foster the abilities claimed to be taught by technology?).

(f) Very small numbers of girls chose to take a technology course in the schools. While this might be largely attributed to previous experience, parental, peer-group and social pressures, media messages, etc, teachers may have discouraged unconsciously the girls from taking the subject, through their actions and speech ("we are trying to attract the sort of boy who . . . "). It is probable that such unconscious messages are read by pupils, and that real changes in the ratio of boys to girls in technology must await a change in teachers' and society's attitudes[2].

A place in the curriculum?

Goodson[3] has analysed the barriers to the inclusion of new curriculum areas/subjects in the already crowded English secondary school curriculum and many of these apply to other countries with differing curriculum traditions.

The lack of a clear, co-ordinated curriculum policy throughout a school presents a real difficulty for any new subject seeking to gain acceptance. Technology does not carry the status of an established academic subject, and there is a lack of consensus as to the nature of technology as a school subject. The ambiguity as to whether it is intended educationally for *all* or vocationally for *some* has hindered its recognition by employers and higher education who tend to rely on well-known parameters like mathematics and physics for selection purposes or on their own testing procedures. Consequently, pupils, parents and teachers are reluctant to give the subject the recognition it deserves. Teachers often perceive science as a body of knowledge and courses as content dominated, while CDT teachers often see their subject as concerned with the mastery of skills. In consequence technology may be perceived as the application of scientific knowledge and technical skills, rather than as a problem-solving process which may draw on the resources of *both* science *and* CDT.

Progress can only be made by the development of a whole school policy for the curriculum, in which technology is clearly defined and has a place in the education of all pupils to age 16. In the early years of secondary education (typically for ages 11–14) technology may be properly introduced as a problem-solving process acting in a societal context, by both science and CDT teachers, operating within their own curriculum areas but communicating to formulate an agreed curriculum avoiding overlap and repetition. Again for the 14–16 age group, given the congested state of the curriculum, it may only be in a minority of schools that separate technology courses, which do not overlap with science and CDT, will flourish. There is anyway a danger that pupils taking a separate technology course will have repetitive experiences. Also it seems very likely that in English co-educational comprehensive schools in the foreseeable future, very few girls will choose to take such a course. Therefore the strengthening of technology *within* the teaching of both science and CDT seems likely to provide a realistic way forward. This will require significant changes of emphasis in the teaching of both subjects and, much more radically, will require tighter prescription on curriculum choice to ensure that both science and CDT are pursued to age 16 by all pupils. There is already a strong movement towards enhancing the technological perspective in the science curriculum, the new policy statement for science containing many positive statements relating to technological problem solving and to the need for good communication between science and CDT:

"Links between teachers of science and CDT teachers are vital if a damaging and unnecessary division between science and technology is to be avoided. All science courses should have a technological content, and all technology courses should have a scientific content. Science courses should help pupils to identify practical possibilities which can be exploited by the application of science in technology. Having identified and specified technological problems, pupils should be given opportunities to develop the art of seeking possible solutions and evaluating them. Hitherto problems tackled in science courses have rarely required pupils to use their scientific knowledge and skills in the design and development of a device or system capable of providing an answer to a particular need. Courses need to be developed which foster the essential scientific content, skills and processes while at the same time providing opportunities for related technological work."[4]

Many teachers of CDT are moving similarly towards incorporating their traditional craft skill course materials into a wider technological framework, although the debate continues as to whether skills in the manipulation of materials must precede problem-solving activities[5]. There is a crucial shortage of teachers of CDT with the breadth of background to take on this new dimension to their work. The number of women CDT teachers is minute and there seems little that can be done to aid recruitment at a time of shrinkage in the teaching force as a whole. One of the crucial needs is for first-class in-service education for teachers of science and CDT to give them the confidence and skills to venture into new areas.

To encourage the teaching of technology within science and CDT curricula, more support material needs to be published for both teachers and pupils, at both national and local levels. The paucity of material relating to technology in schools is very striking when viewed against the array of textbooks and project packages available for any of the regular science subjects. Reay has pointed out[6] that the uptake of a curriculum may owe much more to the ready accessibility of attractive textbooks that to its inherent worth!

To conclude then — some encouraging signs but a long road to travel before a proper technological perspective is to be found in the baggage of all our secondary school leavers!

Acknowledgement

To my colleagues Melanie Nash and Brian Woolnough in the research project "Factors affecting the uptake of technology in schools", but the inaccuracies remain mine.

References

1. Schools Council (from 1980), "Modular Courses in Technology", Edinburgh, Oliver and Boyd. Titles available in the series are — Energy Resources, Electronics, Mechanisms, Structures, Problem Solving, Materials Technology.
2. Nash, M., Allsop, T. and Woolnough, B. E., "Factors affecting the pupil uptake of technology at 14+", *Journal of Reseach in Science and Technological Education*, Vol. 2, No. 1, 1984, pp. 5–19.
3. Goodson, I, (1981), "Becoming an academic subject: patterns of explanation and evolution", *British Journal of Sociology of Education*, Vol. 2, No. 2, pp. 163–180.
4. DES (1985) "Science 5–16: A statement of policy", London, HMSO.
5. HMI (1982), "Technology in schools", London, HMSO.
6. Reay, J. (1977), "Summative evaluation of Caribbean Integrated Science Projects", *New Trends in Integrated Science Teaching*, **4**, Unesco.

9

School Physics and Rural Technology

D. G. SWIFT

Huddersfield Polytechnic, UK
formerly, Appropriate Technology Centre, Kenya

Is it relevant? This is a question that is being increasingly asked about the school curriculum in many parts of the world. Surely nowhere has the dichotomy between school instruction and the pupil's background been more marked than in the teaching of secondary school physics in a developing country. The typical secondary school pupil comes from a rural background, probably growing up in a simple dwelling with no electricity, running water or other services. Average pupils are unlikely to proceed to a higher level of education, or to obtain employment in the modern sector of the economy. The best hope they have of serving their country, and of obtaining a reasonable standard of living, is by contributing to rural development. Even if they do obtain employment in the modern sector, or go on to a higher level of education, they will probably still own land and thereby be able to contribute to rural development. (The rural teacher, for example, as a landowner, is often the first person approached by agricultural extension workers seeking to introduce some modern development into the area.)

These pupils have "traditionally" been taught, either by lecture, use of an imported textbook or by using apparatus that is unfamiliar to them now and which they are unlikely to meet again. They will use the apparatus to measure in a number of ways such quantities as the specific heat capacity of water or metals, the focal length of a lens and the charge/mass ratio of the electron. These concepts and skills have not been presented to them in a way that shows any relevance to their present environment and needs, even from a philosophical, non-materialistic viewpoint, and it is highly unlikely that they would make use of them again in the future. On the other hand, the cost of imparting them has been considerable to the country as a whole, to the local community and to the individual child.

This need not be the case. School physics can be, and in many cases is being, taught as a subject that is seen to be relevant to rural development in an African developing country. This can be done without diluting the mathematical or conceptual level of the physics course, or omitting any of

53

the basic subject matter. In this paper, we will consider an example, both of the problem and its possible solution, from the Kenyan context.

Rural development and school physics in Kenya

Kenya is primarily an agricultural country with more than 90% of its population living outside the main towns and 90% relying on agriculture for their living. More than 70% of these are engaged in subsistence farming. Only 2% of the population are employed in industry and only 3% in the service sector[1]. The overwhelming majority of secondary school pupils come from the rural areas and attend rural schools.

Although Kenya is by no means one of the poorest developing countries in Africa, there are still severe problems. Kenya does not have its own supplies of oil, steel and many other essential materials, and has increasingly to rely on cash-crops to pay for these imports. In 1980, crude petroleum imports used 36% of export earnings, amounting to more than the total earnings from Kenya's single main export: coffee[1]. This drain on foreign exchange reduces the imports available for development and the maintenance of the service sector. It also puts further demand on fertile land already insufficient for Kenya's growing population.

With a net 4% birth-rate, Kenya's rapidly increasing population is putting a strain on many resources. Fertile land amounts to no more than 15% of Kenya's land surface. The settlement in semi-arid small-holdings of those who are unable to obtain fertile land and who are unaccustomed to this new environment is causing a movement towards increased desertification. Another scarce resource is forestry, with trees being felled for firewood and charcoal at ten times the rate if forests are to be conserved. On top of this there is the constantly increasing need for more employment, schools, medical facilities, town housing, pumped water, electricity and other developments.

To alleviate these problems, there is a need to reduce dependence on petroleum-based oil and firewood, to improve water supplies, crop storage and processing, to improve arid-land farming techniques, and to provide cheaper and better housing, and small local businesses that can increase employment opportunities and provide some substitution for imports. This can only be achieved through education, by making the local population, policymakers, engineers and others aware of the needs, and giving them the skills and the understanding necessary for their solution. Because of the technical nature of the problems, education in physics is seen as an important component.

On the other hand, schools themselves are a significant drain on the community. Kenya spends around 25% of its total recurrent expenditure on education[1]. Local fund-raising is needed to assist with the school construction programme Teaching equipment and materials are expensive, especially for physics instruction. Moreover, any school usage of fuel and land for example, further depletes local resources. The payment of fees and other school-related expenses is a major drain on individual resources. Some parents are having to make enormous sacrifices just to keep their children at school for one more term. The hope for the poorer parents is that school will form a way out of the poverty trap for the child, who will later be able to support the rest of the family. The psychological pressures on these children to succeed are tremendous. Because the aim is to get the children away from the "rural poverty" to the urban "pot of gold", the children are increasingly alienated from rural concerns, duties and skills, even when on vacation.

How justifiable then, is school physics instruction in helping to alleviate Kenya's rural development problems? Until recently, Kenya used syllabuses of the Cambridge Overseas Certificate of Education. These were geared to university entrance, emphasising the subject matter needed for those who would read science or engineering at tertiary level. According to the report of the 1976 Kenyan National Committee on Educational Policies and Directives[2], this would be at most 10% of secondary school students, and the figure will not have grown significantly since then. A further 10% directly entered the modern sector of the economy and to these a traditional British physics course might be considered relevant. To the remaining 80%, the physics learnt at school would only be relevant if they could employ it in everyday life, particularly in relation to rural development. This was unlikely to occur unless they had experienced such applications of the subject at school.

The domination of the interests of the "high-flyers" in school physics courses might be justified if the country's problems were exacerbated by a shortage of scientists and technicians. From 1975 statistics, this seems unlikely. For example, whilst there were only 360 posts for scientists and 180 technician posts in research and experimental development work, there were in each case in excess of 5000 people suitably qualified, at least in regard to academic qualifications[3].

It is interesting to note, in passing, that in a 1973 survey[4], only 5% of a sample of 1000 secondary school pupils expected to be involved in agricultural occupations, whilst 40% (rather than the real figure of 8%) expected to become teachers. Another 17% expected to become engineers In all, 90% expected to obtain salaried employment in non-agricultural jobs. In practice, at most 25% obtained such employment[2]. Was this unrealistic and unhelpful bias of pupils' expectations away from agricultural and towards modern-sector, salaried employment partly a result of the non-agricultural bias of school physics?

It is perhaps hardly surprising that the National Committee's report was critical of the lack of relevance of school science syllabi to the country's needs, particularly in regard to rural development. They recommended the government "to give prominence to the teaching of agricultural sciences in secondary schools and to relate the teaching of other subjects to agriculture", and also "to localise science content and methodology as a basis for development of labour-intensive technology appropriate to the support of basic activities of life in the rural areas".

Is physics relevant?

Is it realistic to expect that school physics could be made relevant to rural development?

A simple way to obtain a positive answer is to define physics in terms of its process skills rather than its content. Pupils who have trained as scientists, who have been taught to observe carefully, take and record accurate measurements, process and display these in an easily digestible form, and make appropriate generalisations, are going to be better able to tackle problems in any field, rural development included. This was the rationale behind the East African Nuffield/PSSC-style "Secondary Science Project". However, even this contained an important bias towards local applications. One example was a locally-biased section on Materials and Structures. Another was the build-up to considerations of national Energy issues rather than a concentration on the electron that formed the climax of the parent Nuffield scheme. Unfortunately, the emphasis on expensive and time-consuming pupil experiments was unacceptable for the average Kenyan school with lower finances and a shorter course (4 years rather than 5 years) compared with the UK. The SSP scheme has therefore now been superseded.

Even with a "traditional", content-oriented course, there is still ample opportunity for increasing relevance. (I have described elsewhere[5] how school physics is very relevant to rural development, and will only give a few examples here.) An appreciation of forces, density, pressure and centre-of-gravity is needed for a full understanding of the design of dams, water towers, farm structures, and even the pruning of bushes and trees. Dynamics and fluid dynamics are needed for the design of water-supplies and pumps. Heat flow — conduction, convection, radiation and evaporation — are important for the improvement of wood-burning and charcoal-burning stoves, and for the design of home-made solar crop-dryers and solar heaters, all of these being ways of reducing the current drain on firewood. Electrolytic action is important for rust prevention in water-tanks and water-systems. The list is endless. Much of school physics is as relevant to the concerns of rural development as it is to those of an industrialised society. The problem is to teach it in a way that brings out this relevance.

Ways of teaching physics for rural developments

The traditional way of introducing applications of physics is to tag them on at the end; to briefly discuss them as a means of showing the practical importance of concepts that have already been thoroughly analysed. This has at least two disadvantages — the relative emphasis on the importance of the physics itself rather than the applications, and the lack of time — and stamina — available for considering the applications.

A better way, perhaps, is to use practical environmental concerns, familiar to the pupils, as "problems" requiring the use of physical concepts. In other words the environmental concerns are used to introduce the physics topics, thereby providing better motivation and better understanding by relating them to the pupils' experience. Such "problems" also serve as a better way of testing understanding of the underlying physics[5]. The "problem" can be expressed in practical terms as a student project, or used as context for a written examination question.

In the ideal situation without severe time and other constraints, the "problem" could form the basis of a long project, possibly involving visits (say to a building site near the school, or the school's water or electricity supply) and out-of-school work, perhaps through the science club. For example, the pupils could investigate the possibility of installing a biogas plant or hydraulic-ram water pump at the school. They could take measurements, and work out a suitable design. They could, perhaps, even obtain funds from a development agency or government department to obtain the equipment, and assist in its installation and subsequent maintenance. Apart from the obvious benefits to the pupils, this project could in the long run guarantee a well-maintained school biogas or water supply (poor maintenance being a major source of biogas problems in Kenya), save on school fuel costs, reduce the pressure on other scarce fuel resources, and act as an ideal demonstration to others in the local community.

In a sense, such a biogas plant or hydraulic ram would be acting as a piece of science apparatus. This illustrates one other easy way in which local relevance can be improved. Wherever possible and sensible, equipment, materials and situations that are familiar to the students should be used, rather than giving the pupils the impression that physics is something that only happens in a laboratory.

Making a start

Several other initiatives can be taken on a national scale, and have been taken in Kenya leading to an increase in the relevance of school physics.

One notable example is the preparation and introduction of new syllabi. The new Certificate of Education course seeks to combine the best of the "traditional" (i.e. Cambridge-style) and SSP courses, with increased emphasis on local relevance. The first two aims of the course are to help students to understand the world around them, and to make students aware of the effects of scientific discoveries and knowledge on everyday life through some examples of applications of physics. The syllabus contains, for example, reference to the efficiency of the local charcoal stove, solar-heaters, sources of energy including biogas, fibre and rod reinforcement applied to concrete and to sisal, and several other explicit references to rural-development concerns.

Secondly, the National Examinations Council has laid an emphasis on rural development as a suitable context for examination questions in physics. This is evident in some of the recent examination papers, one, for example, placing a question on density in the context of the mixing of concrete.

Another positicve contribution is the setting up of an Appropriate Technology Centre at Kenyatta University College which houses the Education Faculty of Nairobi University. This Centre is, amongst other things, providing rural-development projects courses, and Energy and Materials courses for the Physics Department, plus an Environmental Education course for a cross-section of Education students, and in-service courses for school-teachers.

Even before these developments, the Inspectorate had encouraged teachers, including physics teachers, to lay greater emphasis on the environmental aspects of their subject, with examples being given to assist the teacher.

Finally, there is the long-standing "Student Science Congress" competition that encourages pupils, usually in school science clubs, to develop physics and other science projects of relevance to rural development.

Thus it can be seen that Physics is relevant to rural development, and can be taught in a manner that makes this relevance evident. Moreover, Kenya (for example) has already taken steps to reduce the dichotomy that had previously existed between school physics and rural development.

References

1. *Europa Yearbook 1982*, Vol. 2, London 1982.
2. *Report of the National Committee on Educational Objectives and Policies*, Nairobi, 1976.
3. *1981 Statistical Yearbook*, United Nations, New York, 1983.
4. Keller E. in *What Government Does*, Vol. 1, ed. Holden M. and Dresang D., Sage Publications, Beverly Hills, 1974.
5. Swift, D. G., *Physics for Rural Development: A Sourcebook for Teachers and Extension Workers in Developing Countries*, Wiley, Chichester, 1983.

10

Assimilation of Technology in Rural India — An Educational Approach

S. R. MEHTA
Vikram Sarabhai Community Science Centre, Navrangpura, Ahmedabad, India

A vivid example of the application of science and technology in our lives in India is the National Biogas programme, a programme which is making a large scale intervention into our traditional rural society.

As a result, millions of rural people will come into direct contact with a new technology which will confront them with scientific ideas in opposition to long-held beliefs and prejudices. This is a challenge to those who are developing science and technology curricula for schools. However, the programme itself can become one of the examples of the ways in which we can educate our future citizens.

As a first step, we have broken down the scientific and technological principles involved in the building and use of a biogas unit — in terms of the Gujarat State Higher Secondary School Examination.

Further, we are developing simple experiments to produce biogas and examine its properties and we wish it, too, to have a technical bias so that students gain experience in simple work with wood, metal and bricks in making a unit. This is one example in a programme for introducing rural technology into our schools and showing its relevance to today's needs.

11
Rural Technology for the Landless in Bangladesh

M. IBRAHIM
Centre for Mass Education in Science, Dhaka, Bangladesh

In Bangladesh, as in many other developing countries, knowledge of appropriate rural technology has not, so far, benefited the rural landless in any significant way. To translate the technology into action for these highly disadvantaged groups something more than just availability is involved. It needs an appropriate education. A voluntary organisation in Bangladesh, the Centre for Mass Education in Science, started a project which is taking some experimental steps in that direction.

The target group

A majority of the rural population of Bangladesh consists of the landless poor with little or no agricultural land of its own. This forms the most disadvantaged segment of the society, although many within it have the basic skill and energy to provide various important services to society. Women form a particularly neglected group whose potential as contributors, to themselves and to the society at large, has never been appreciated, least of all by themselves. Being brought up in an exploited and hopeless circumstance the children from these groups are condemned to perpetuate the situation without any education, and any chance of a breakthrough. As for education, a lot is being talked about "Universal Literacy" in this land where only 22% of the population (12% of the female population) is literate. If we offer to teach only how to read and write to a group whose immediate problem is that of keeping alive, the offer may not be well received. We have to demonstrate that education can be a powerful ally in one's struggle for a better life, and even for survival.

The experience of Grameen Bank (an innovative banking system for the landless in rural Bangladesh) shows that these very groups can be electrified into all sorts of productive activities through skill and labour when a start is made by providing a small bank loan. The results with the women were specially appealing. But the landless, in the confines of their situation, soon exhaust their potential because of the lack of proper technology. They will need an appropriate but better technology to move from the present potential to a higher one, and an appropriate education should be able to generate such a higher potential in the shortest time. The content and the method of education would be such that this would make life-oriented knowledge and technology available to the disadvantaged target groups at close quarters so that the adoption of the technology becomes an instant reality, not a mere future possibility.

Examples of rural technology

When we talk about rural technology, we must remember the perspectives. Many of them sound so commonplace, yet they are so vital in the improvement of life for the rural multitude. Let us make a short list.

1. Health and family planning

(a) Pure water: purification, tubewell and other sources.
(b) Nutrition: choice of food, cheap but nutritious foods.
(c) Hygenic home: materials to build homes, latrines, sewerage, pest control.
(d) Common maladies: simple treatment, preventive and curative.
(e) Family planning: motivation, methods and materials.
(f) Child and mother care.

2. Food processing and preservation

(a) Processing of paddy: improved methods for threshing, drying, husking, parboiling, etc.
(b) Storage of paddy and other grains.
(c) Preservation of vegetables, fruits, fish by solar drying, salting, pickling, etc.
(d) Making of bur and brown sugar.
(e) Improved cooking, cookers (chula), solar cookers, hay-box.

3. Cattle, goat, poultry raising: better feeds for them, better management.
4. Agricultural and horticultural methods and techniques.
5. Fish culture including small ditches.
6. Seri-culture, eri-culture and api-culture.
7. Textile: weaving, natural and chemical dyes, block print, screen print, etc.
8. Maintenance of machineries: e.g. tubewell, shallow power pumps, rickshaw, auto-rickshaw, etc.

9. Metal crafts: blacksmithy, tinsmithy, making of small spares for simple machineries, workshop.
10. Taking advantage of rural electrification programme.

We may add many more things to the list, but this gives a sufficient indication.

An experimental rural technical school

The centre for Mass Education in Science started an experimental rural technical school in the village Suruj in Tangail District in 1981. This has been a modest grass-roots establishment to provide disadvantaged boys and girls with a basic education for improving the quality of their lives while learning and practising a trade useful to the community to become an income-earning member of the family. The school is also meant to serve the community as a rural technology centre by undertaking repairs, maintenance, fabrication and such other things, as well as training. A bamboo-tin structure on donated land, the school campus is homogeneous with the surrounding rural scene with its trees, vegetable garden, mini-pond and make-shift workshop. Much of the furniture and facilities in the school are actually made there, some by the students themselves. The timing and mode of work in the school were developed through trial and error to suit the requirements of the target group.

Some important aspects of the school are:

(i)   A curriculum relevant to skill development and of immediate use to the rural life and productivity. This is a condensed package preparing the students to take their place as income-earning members in the community.
(ii)   Education is not a luxury here. It is flexible enough to meet the usual routine and the requirements of the rural disadvantaged groups.
(iii) There is as little gap as possible between education and its practice in real life. A good part of the education consists of such practices and the transition to the income-earning situation is a smooth one.
(iv)   The school serves as a service and resource centre for rural technology and the students act as the agents for the improvement of life through the extension of such technology.

The students are offered a condensed life-oriented general education as well as a training in some trades. The curriculum for the former includes: use of literacy; arithmetic; measurement and accounting; technical drawing; plants; health; materials; energy; machines; country; and society. Each one of these subjects involves practical work, field work, and project work made as realistic as possible. In fact many of the classes take place in appropriate places instead of the class room, for example, plants are studied in the vegetable garden. Here are some examples of practical work that go with the teaching.

> For Arithmetic and Measurement: land measurement, estimation of materials required to make a piece of furniture or a dress, household accounting and accounting for the activities in school.
> For Plant: planting and observing vegetables, trees, experimentation with fertilisers, making compost etc.
> For Country: collecting information about the village (land, water, agriculture, technology etc) through observation and discussions; making maps of the locality; survey on drinking water, irrigation, fish culture, cattle, poultry, etc.

The trade training includes carpentry, metalcraft, repair of machines and small engines, electrical wiring and installations, tailoring, crop and food processing, api-culture, seri-culture, natural dyeing and block printing of textiles. As training in these subjects went on, the school performed various paid jobs for the community. These include:

(a) Routine maintenance and repair of all hand-tubewells in the locality.
(b) Maintenance and repair of most of the shallow power pumps (about 40) in the area.
(c) Various repair jobs on metal and wooden gadgets; also on bicycles, rickshaws, auto-rickshaws, etc.
(d) Sale of the products of the school like dresses, textile prints, wooden or metallic goods etc, as well as some innovative items like improved cookers and improved lamps
(e) Electrical installation and repair (a part of the area in the vicinity is electrified by Rural Electrification Board).

These activities give a regular income to the school and students. Moreover, the students are already well acquainted with the realities of the market while they are still in the school and learning. The school also serves as an extension centre for the community. As for example, it demonstrates the management of a mini-pond and gives technical assistance to mini-ponds like fish cultivation efforts within the target group. The school actively propagates the plantation of Ipil ipil, the wonder tree; the use of improved cookers; bee-keeping in the household; and such other activities through the students, their families and others in the target group.

A rural technology project

The experience gained from the rural technical school has resulted in an expanded project namely "Rural Technology for the Landless: Education and Extension through women and children". This aims to:

(a) Explore the available rural technology and to educate the target group in its adaptation by providing necessary technical support;
(b) Arrange for a basic life-oriented education with emphasis on technical skills of immediate income-earning potential within the community;
(c) Devise and test the content and the method of an appropriate education for those who are being left out of the education system;
(d) Arrange for the dissemination of the experience gained by the project.

The programme is executed through several Rural Technical Centres and schools, similar to the one described above, and a central "Service Centre". The former are situated in various regions of the country where the Grameen Bank is active. The Service Centre is the supporting facility providing necessary research, planning, materials development, management and dissemination. The Service Centre works in close contact with appropriate technology organisations and groups, and has facilities to replicate and experiment with some of the technology more immediately adaptable for the target group. It helps the schools in the transfer and adaptation of such technology.

At the end of its first year, the project is on its way to achieving some of its goals. Apart from the benefits it is bringing to the target group in the immediate villages concerned, the project is expected to have a broader impact. This is an experiment in mass-education aimed at achieving a university of education in the real sense. The experience gained from this experiment and the materials prepared through this may be useful to other attempts, and even to bigger national programmes concerning mass-oriented education and community education.

On the technology side, the project should offer a good testing ground for the real-life appropriateness of some of the technology. As this training is integrated with education and the life of the target group, it has a better chance to be "transferred". Once it is successfully done, the dissemination mechanism within the project will try to let it be known as widely as possible. The project can also contribute to the evolution of proper back-up facilities for the technology which is gradually, but inevitably, entering our rural scene.

The results so far obtained from the various aspects of the experimentation are being disseminated by the project. It has published various booklets and organises discussion meetings and seminars to involve other groups and individuals who are concerned with appropriate education and appropriate technology for the rural poor of Bangladesh. A monthly periodical and a small printing press is helping the project in its dissemination activity.

12

Education and Training for Biotechnology

R. A. KILLE

University of Edinburgh, UK

An attempt will be made here to highlight those areas of the educational and training systems which will need to be reviewed if the schools and universities are to turn out people who are appropriately trained to develop the potential that biotechnology offers.

Biotechnology is fast becoming a vogue. There is an ever increasing number of biologically related processes and industries which wish to be considered as biotechnology. There is a danger that if all technical achievements which have a biological basis are included within biotechnology, it will lead to a loss of direction and thrust in research and training for the wholly new techniques, and perhaps a dilution of the financial support which they will need. We are familiar with the educational and training requirements of the older, more traditional industries using biological systems such as farming, fisheries, forestry and brewing. We need to identify the skills and activities the new industries will require and to examine ways in which educational systems can be adapted to meet the new requirements. In doing so, however, care must be taken to avoid distorting the curricula and the distribution of educational resources as cell and environmental biology and other fashions have done in the past.

The following definition of biotechnology is adopted here: the application of biological organisms, systems and processes to manufacturing and service industries. In particular it is the expansion in the industrial use of microbial and other cells, together with the demands these new manufacturing processes will place upon close integration and understanding between biologists, chemists and engineers.

The use of biological processes in industry is not new. They have been used in the food, drink, pharmaceutical and effluent treatment industries. However, recent advances in cellular biology, molecular biology and genetic engineering, together with new developments in biochemical process engineering and in control engineering, have created completely new industrial prospects. These new industries can be expected to play a substantial role in the provision of better drugs, vaccines, hormones and antibiotics, cheaper home-based supplies of energy and chemical feed-stocks, and improved environmental control and waste management.

Educational programmes must not only provide the necessary training for those who will be involved in these new industrial developments, but they must also aim to increase public awareness of their potential and counter any exaggeration of their hazards. Unlike earlier fashions in biology in which the thrust has been largely in one particular area (such as cell biology), biotechnology will involve several areas of biological science allied to chemical engineering and commercial marketing operations within a particular industry. It will not therefore be sufficient simply to increase the emphasis in a curriculum in a particular subject area as was all too often the case for cell and environmental biology at the height of their fashion. These curricula changes produced biologists perhaps better informed in a fashionable aspect of biology than previously, but largely limited to purely academic objectives. There was often the hope that the new knowledge would be applied to social and industrial projects, but very little if any social or industrial relevance or training was included in the educational programme. Biotechnology will require a much broader understanding of biological and physical sciences, which goes beyond their purely academic interest towards their industrial applications and management.

The educational and training requirements of biotechnology will necessitate a considerable change in attitude amongst teachers in schools, colleges and universities, students and their parents. In many parts of the world, including the developing countries, there has been far too much emphasis on the merits of the honours degree as a qualification for a successful career. Biotechnology will require a limited number of people trained in this way, but the greater number of posts is likely to be in the professional technologist and technician grades. The advent of biotechnology should provide the stimulus and the challenge to educators in schools, colleges and universities and the ministeries of education to provide courses which have a real social and industrial relevance, and which therefore will be attractive to a much wider range of interests within student/pupil populations other than the purely academic. There is no reason to fear that such curricula should be any less rigorous than those used at present, and indeed the level of numeracy alone that the new industries will demand of biologists may well increase the rigour.

Is biotechnology, as defined here, of relevance to developing countries? Biotechnology will not be labour intensive, and it will require in most cases, heavy capital investment. Each country will decide for itself if any of the new biological industries can make a significant contribution to its social and economic development. Certain applications of biotechnology may well indirectly have a beneficial effect on the development of labour intensive industries which are of importance to developing countries. The production of cheap fuels from fermentation processes, fertilisers and protein feed-stuffs are examples of products of biotechnology which could assist the social and economic development of the poorer countries.

Educational objectives

The educational requirements of biotechnology can best be assessed by considering the activities associated with a new industry. These are research, development, production control and marketing. The educational and training programmes of schools, technical colleges and universities must between them take into account these four career outlets in biotechnology.

1. Research.

People with post-graduate training who will provide the basic ideas, techniques and materials for industrial innovation from their scientific and engineering research.

2. Development.

People with good honours degrees or diplomas in technology who will translate new ideas into manufacturing processes and into economically viable products and services.

3. Production Control.

People with degrees or technical training who will monitor and control the manufacturing processes.

4. Marketing.

People with a wide range of training and qualification in financial, sales and management roles in industry; those interested in investment policy, political decision making, advertising, and public relations.

Each group will play a key role in determining the rate of development, acceptability and the social and economic return from the application of new biological techniques in industry. In each group there will be a need for individuals who have acquired an understanding of microbiology, genetic engineering, enzymology, and biochemistry.

Given the wide range of skills biotechnology will demand, and the wide range of potential products and services the new industry will offer, it would seem that except in a few cases, special courses in biotechnology are neither necessary nor desirable. It is doubtful if suitable courses could be devised to usefully serve the development of the wide ranging potential biotechnology offers at the present time. The introduction of such courses would impose a heavy drain on the existing resources of schools and universities. It is unlikely that adequate numbers of staff are available with the necessary scientific and industrial experience. Moreover, the school and university curricula must be continued to be sufficiently well balanced to serve the existing more traditional professions and industries. To fulfil all those roles, the main objective must be on turning out individuals from school and university with a broadly balanced background in physical and biological sciences, with an appreciation of the roles of the engineering, management and marketing in industry. This will require not so much the introduction of new subjects into already overloaded curricula, but a re-organisation of existing educational resources, and above all a change in attitude on the part of teachers, students and parents towards an awareness of the industrial potential of biological sciences, and the career opportunities for professional technologists, technicians and business skills.

How are biologists in educational institutions at the school, technical college and university level, to adapt to the requirements of biotechnology and retain a curriculum suitable for other careers?

Schools

(a) At junior secondary level, it would seem necessary to maintain a broad spectrum of basic sciences, namely biology, chemistry, physics and mathematics. It is possible to introduce into school courses at this level an awareness of some of the application of biological science in industry; some of the problems facing engineers in plant using biological materials; the great importance of numeracy for biologists? There will be great difficulty in introducing some or all of these topics, but they should be given serious consideration. Co-operation between biologists, chemists and mathematicians should make it possible to create at least an awareness of the industrial potential of biology. It is vital that biotechnology is always included as one aspect of biology.

(b) In senior secondary school curricula the educational requirements of biotechnology will exert greater pressure on the curriculum and its assessment. These pre-university years are dominated in many countries by the assessment of the academic abilities of the pupils. It will require a co-ordinated effort and goodwill on the part of the universities and teachers of the final school years to devise and accept the changes that biotechnology requires if it is to be successfully exploited. Teachers will need to ensure that pupils receive a balanced view of the opportunities for employment in biological industries, and alternatives to degree courses must be made clear to pupils. Some would argue that this change in attitude has long been overdue. The advent of biotechnology has merely made it more imperative, for many school syllabuses reflect areas of interest, approaches and skills that were in demand some 20 or more years ago. There must now be selectivity to ensure that syllabuses teach both key scientific principles and generate the necessary awareness and skills for future scientists and engineers.

(c) Such is the speed with which biotechnology is developing, a reliance upon pre-service education and training of teachers will be too slow to provide an adequate number of teachers equipped to meet the requirements of biotechnology in the immediate future, although it is imperative that biotechnology education features in such courses.

To encourage quicker change ways will have to be devised for introducing major in-service programmes of refresher and re-orientation courses. In addition to programmes of lectures and practical work, there will need to be direct industrial contact to give teachers some understanding of industrial and business problems. Liaison between industry and the school will need to be strengthened with the help of the professional bodies, industrial organisations, the trade unions and the college of education and technology.

There is already an urgent need for suitable textbooks, visual-aids, laboratory exercises and suitable biological material, especially in the fields of microbiology, simple genetic engineering and biochemistry.

Technical education

The largest opportunity for a career in biotechnology will come from the technical support it will require. It is also likely to provide job opportunities for a wide range of skills and abilities. Technicians and technical assistants, who will be required to take part in the development, control and monitoring of industrial biotechnological processes, are likely to outnumber four or five times the professional biotechnologists with an honours degree. It is therefore vitally important that, where necessary, steps are taken to raise the status of technicians and to persuade students of average performance (and their parents) of the merit and career potential of courses other than those leading to a university honours degree.

The curricula of most technical colleges will probably require very little modification to provide the scientific and industrial training requirements for biotechnology. It may be desirable to create one or two colleges of biotechnology which concentrate their training on a limited number of industries. Most colleges should perhaps aim to retain flexibility and adaptability by providing courses which concentrate on a firm foundation in technique and skills of those subjects which contribute to biotechnology, leaving the specialist training to the training schemes of individual industries.

Consideration will need to be given to ways in which technical colleges can contribute to the re-orientation and refresher training of staff in schools and other institutions, and those already in employment, by way of sandwich, day-release or other systems. In countries where the development of industries in biotechnology is rapidly taking place, there may well be a big demand for this function of technical colleges — to provide sufficient numbers of technologists to control and monitor the plants.

Universities and colleges of technology

These tertiary level institutions have a major role to play in the development of biotechnology.

It is in the universities and colleges of technology that the professional technologists will be trained. The high fliers with good honours degrees, or the equivalent diploma, and a postgraduate qualification, will provide the new ideas for the use of biological systems from their research. Those with reasonable honours degrees will contribute to the development of these small-scale laboratory systems into industrial processes.

Universities differ considerably in their course structure, number of subject departments and overall size. Individually they will have to consider a number of issues. Under what conditions will it be advantageous for universities to pool their resources in staff and equipment with those of local industries in a biotechnology centre? What conditions will make a course in biotechnology *per se*, within a college or university a sound policy decision? Most large universities will probably choose to incorporate aspects of biotechnology into their existing course structure. This would enable them to continue to offer a wide range of professional training options and incorporate biotechnology with the minimum of extra resources and disruption of their course organisation. This system would also have the advantage of a flexible response to developments in biotechnology through their ability to produce graduates specialising in a wide range of subjects involved in biotechnology, such as biochemistry, molecular biology, microbiology, genetics, plant and animal cell biology, chemical engineering, process control engineers, food processing engineers and business studies.

Whatever strategy is adopted there are educational objectives which will always have to be borne in mind. The applications of a biological science will have to be given more prominence than it has been given heretofore. As well as the constraints on biological systems in manufacturing plant, students should also be made aware of the physical and engineering problems. This may mean the provision of engineering courses for biologists, and in return the biologists could with advantage provide courses for chemical and process engineers. Where the degree curriculum is a modular system there should be little difficulty in arranging a suitable coverage of the essential biological discipline, including a greater acceptance of numeracy as a basic qualification for biologists and some awareness, of engineering and marketing practice.

The incorporation of aspects of biotechnology into subject specialisations carries the risk that the applied side will be treated superficially, especially if the task is given to a staff member with no industrial experience. The provision of staff with suitable understanding and experience of manufacturing industry will be one of the problem universities will face. Initially this may be overcome by guest lecturers from industry. Longer term it may be necessary to link academic staff closely with industries on joint research and development projects through sabbatical secondment.

Since new uses for biological systems are most likely to come from fundamental academic research, there would seem to be little need to change the methods of training of postgraduates. At most, more funds and facilities should be made available for postgraduate studentships in promising fields in microbiology, genetics and biochemistry.

However, one or two year M.Sc. — type courses in specific areas of biotechnology for newly qualified graduates may be necessary. In those countries where there is to be a rapid expansion in biotechnology consideration will need to be given to ways in which colleges of technology can provide re-orientation and refresher training in day-release or sandwich course systems for those postgraduates already in employment. There may also be a demand for such courses from those people whose task will be to control and monitor manufacturing plant.

Finally, the universities are in a strong position to influence the attitude in schools to industry and commerce. By modifying their entrance requirements they can make it easier for the schools to introduce applied aspects of biological science into their teaching and assessment. The universities are also in a strong position to assist teachers, parents and pupils towards a better appreciation of the potential for the new biological industries.

Community education

The exploitation of biotechnology for maximum social and economic development will depend to a great extent on changing the attitudes of a wide spectrum of the population.

Biologists must accept the task of persuading and assisting the newspapers, television and radio in educating the general public about biotechnology, its need for practical skills and the commercial opportunities they will provide. Such publicity would be directed towards creating an awareness of the opportunities for those not academically orientated, and hopefully encourage a demand for changes in the schools.

Publicity for biotechnology in the media would need to be re-inforced by talks, demonstrations, exhibitions and seminars to give more detailed technical information of the requirements of biotechnology to those in positions of influence and decision making, such as education officers, civil servants, industrialists and financiers.

Acknowledgement

The author acknowledges that a number of ideas in this paper are derived from the report of a Working Group of the Royal Society on "Biotechnology and Education", London 1984.

42   DISCUSSION

This sequence can go either way as teachers gain confidence. The help which teachers need is suggestions of real problems which could be used, along with appropriate teaching materials. Once again, in-service training is important.

All the approaches above are variations of the "Science-first" approach. If 'Applications-first' is likely to be more effective in teaching about the science and technology which will dominate the lives of our students, will we have long to wait?

13

Teaching Biotechnology

K. JOHNSEN

Rungsted Statsskole, Rungsted Kyst, Denmark

During the seventies and the eighties technology has played an increasing role as a teaching subject in the secondary schools of Denmark. The reason for this is obviously related to the pressure from the environmental and economic crises in the Western World. The rapid increase in existing technology and the expansion of new technologies call for rapid changes in education. This paper concentrates on the possibilities and problems of teaching technology in biology at the secondary school level. It is important to emphasise that technology teaching ought to be interdisciplinary, but that the structure of the Danish school system makes this impossible.

Biology teaching in the secondary schools of Denmark

In Denmark biology is taught at two levels in the secondary schools: three weekly lessons in the final school year of the lower secondary school, and at the higher level, three weekly lessons in the second year and seven weekly lessons in the third and last year. At both levels the purpose of the education is to "improve the student's ability to formulate and solve biological problems" and to "comprehend biological methods and reasoning in a critical and analytical manner", and furthermore to "use biological knowledge and methods on individual as well as social problems". The curriculum is not described in fine detail, giving students and teacher a choice of subjects within certain limits.

A theoretical framework

In order to plan instruction in technology it is necessary to have a good framework of understanding. The model presented in Fig. 1 is suitable for several reasons. First, it places technological improvements in different sections of the production process (objects of labour, means of labour, and labour power). Secondly, the model does not only focus on the production process, but also on the surroundings (the environment and the human society). Thirdly, the model can be viewed in a dynamic way, the different parts changing in time and space. Changes in one part of the model give direct or indirect changes in the other parts — sooner or later. For example,

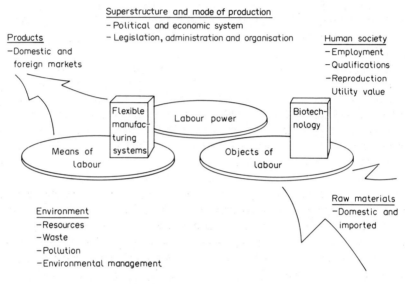

FIG. 1

production can change the environment directly either by pollution or by depleting resources. When pollution reaches high levels, individuals react by forming movements — or the established society reacts with restrictions. In this way the environment indirectly affects production and the technology. New objectives may require new means, new means may require new qualifications for the labour force. A new form of production can directly prevent another, thus forcing change.

New technologies are part of the model as they require flexible manufacturing systems, and new means of production require quite new qualifications in the labour force.

The purpose of secondary school is to teach the students not how to use technology, but to evaluate technology on the basis of a general awareness of how the technology is used and how it affects society and the environment. In addition students should have specific knowledge of technological methods and principles as they affect further education and future employment.

Such a model is too abstract for most students and practical work should therefore always be integrated in technology teaching as a basis for theoretical discussions as shown in the example below.

Traditional technology

The following example shows how the theoretical model can be used in practice. My students wanted insight into food-processing technology through a practical example of food production. A serious accident at a factory in Copenhagen producing vegetable oils and soya-bean cakes was the motivation for working with margarine. The accident was caused by the explosion of benzene. The first questions from the students were: Why is benzene used in the production of vegetable oils? Why are people talking about the dangers of chlorine? Why all this talk of mercury?

They decided to work out a flow-diagram for the production process — it was in fact possible just from the newspaper articles (Fig. 2). The students found it exciting to look into a new world, and they decided to make margarine from soya-beans step by step in order to get a better understanding of the processes in the factory.

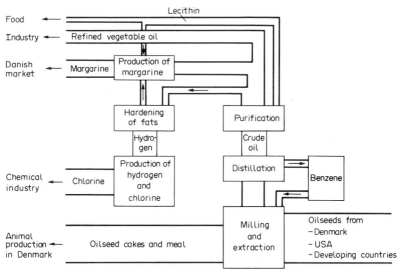

FIG. 2

Besides getting an understanding of food technology, the students along the way posed a number of questions that are necessary for the evaluation of technologies. The questions can be categorised according to the model (Fig. 1) starting on the right side of the figure.

The *first* category of questions concerned the raw materials and their origin. The questions gave rise to discussions on the protein import from developing countries, on the international market for nutrients, and on the historical background for the location of the factory, which lies at the harbour in the centre of Copenhagen.

The *second* category of questions concerned the production itself, the means and processes. These were answered through practical work in the laboratory. The students who had almost no experience in laboratory work learned a lot about the properties of lipids and proteins, about purification, distillation, separation, and centrifugation processes, etc. They also learned how to handle various items of equipment, observance of safety rules to avoid hazards (handling strong acids and bases, mercury, chlorine and the distillation of benzene). This of course gave rise to the *third* category of questions concerning the work environment in the factory. In particular the problems of organic solvents affecting the nervous system was discussed.

The *fourth* category of questions concerning the outer environment came up naturally as hydrogen was produced with a mercury electrode and chlorine as a secondary product. The uneasiness of the local population became understandable. Later on in the course this problem was further analysed in a project concerning mercury in the environment.

Finally the role of the products in the Danish society was discussed, particularly the nutritional value of the produced margarine and the role of food additives in provoking allergic responses. (The students actually produced margarine from oil by adding hardened fats, water, milkpowder, colour, emulsifier, and salt, but the lack of a usable flavour make it extremely distasteful).

Biotechnology

Biotechnology plays a major role in the new technologies. The term "biotechnology" includes a number of technologies, originating from molecular biology, and usually defined as "the technical use of biological knowledge in production". In this broad definition the bread-making process with the use of yeast, alcohol fermentation, selection of domestic animals, etc. are included. In the context of "new technologies" the driving force is comprehension of *biological information*, and only the "deliberate use of biological information in production and reproduction" will be considered as "biotechnologies" in the following.

Biological information is stored in different macro-molecules, but until now attention has been focused on DNA, RNA and proteins — mainly because information for other molecules was not accessible. Different techniques have been developed more or less independently. Protein production from micro-organisms for specific purposes has long been known, but only when combined with the techniques of splitting, splicing, and cloning DNA from various sources did it become a revolutionary technical improvement.

In industrial production

The use of biotechnology in industrial production is based on the fermentation technique — alcohol fermentation being an old and well-known example (fermentation is the microbial production of specific products by monoclonal cultivation under optimal conditions. The products are more or less effectively purified from the nutritional medium). Through a better understanding of mutagenesis and selection, the fermentation technique has been improved by making the micro-organisms still more effective in producing the desired products. The increased application of enzymes and other fermentation products has also resulted in a rapidly growing market. The introduction of immobilised enzymes has made it increasingly economic to use enzyme techniques instead of traditional chemical processes. Furthermore enzymes provide many advantages in terms of energy use, intermediate and waste products being biologically convertible, creating less serious hazards, etc.

With the introduction of genetic engineering the number of different products made by fermentation techniques has increased considerably. The creation of artificial DNA is only limited by our understanding and imagination. As an example the development of cheap immunoglobulins could revolutionise purification processes.

In agriculture

Selection of plants and animals with desired characteristics is an ancient technique, but has until recently been a very slow process. The freezing of semen, eggs and embryos offers new prospects. Hormone-induced superovulation and *in vitro* fertilisation increases the speed and possibilities. The success of inserting specific genes in fertilised eggs can induce revolutionary changes in the procedure — thus it is not only characters originated by chance that can be selected, but also characters introduced from other organisms.

In plants the efforts are concentrated on producing species which require smaller amounts of fertilisers and pesticides. Transfer of genes between species opens up the possibility of making nitrogen-fixing cereals. These techniques introduce several positive aspects including more economical use of organic production, reduced need of fertilisers and pesticides — resulting in more food for the world population and a cleaner environment. But it may also lead to the depletion of the total genetic information in the world leaving fewer possibilities for future generations. And one should not forget the side-effects of the so-called "green revolution" introduced during the sixties in the developing countries.

In medical treatment

Many medical products are made by fermentation, but until recently they have been limited to naturally occurring microbial products, although

EIT-D

some chemical modifications are done. Introduction of genetic engineering removes this limitation, and in principle all naturally occurring products can be made by micro-organisms. It should be stressed though that the question of making species-foreign products is not so much a question of finding the right piece of DNA, but more of finding the right conditions for the production.

Within the last few years the making of "test-tube" babies has become routine work, and the techniques evolved in agricultural research on animals are applicable to humans — with all the moral and ethical implications. The rapidly growing understanding of human genetics and chromosomes in combination with immunological and enzymological techniques has already given prenatal diagnostics a push forward.

Chimera probably are of little economic interest, but they have great scientific importance in studies of human genetics and development. Chimera of humans and chimpanzees in test-tubes might facilitate the way to better understanding of communication between cells and genes. Gene therapy has been practised on two occasions, but was strongly critised by experts. Not because of the ethical implications — but because the techniques have not yet been sufficiently developed.

In the future

Many of the biotechnological inventions are still on the laboratory scale, but the investments in research are so huge that the question of practical use is largely a question of time. A glimpse at the stock market gives the assurance that biotechnology is likely to succeed due to the size of investments alone. An example are the investments in research of the pharmaceutical concern NOVO comparable to the entire budget of the Faculty of Sciences at the University of Copenhagen.

A short official note from the Common Market requests that the member countries should make efforts to strengthen the biotechnological research in order not to lose ground to countries such as the USA and Japan. The note predicts the coming of a "bio-society" succeeding the "information-society". At least two new areas of great interest are in the making: energy-production and the "bio-computer".

Conversion of solar energy to electrical power by plants and the splitting of water to hydrogen and oxygen are relatively well-known processes. The technical use of that knowledge could give several advantages: reduced problems with pollution and resources and a lower demand of materials for installations.

The code of DNA is also well known, although more knowledge tends to complicate matters. The information of proteins is intricate and rather poorly understood, but possibilities seem to be innumerable. The "protein-chip" might not only be advantageous in size and energy spending, but in function as well, making even the smallest chip of the eighties look coarse and primitive.

To the layman both examples are rather speculative, and questions of reproduction techniques are much more urgent. How far do we want the development of prenatal diagnosis to go? What are the juridicial and ethical implications of the "test-tube" techniques? And so on.

The Danish Home Secretary has made an effort to describe these problems — unfortunately in a very inadequate report. In the society there is a tendency to a dual development: on one hand a continuous introduction of new techniques in hospitals, on the other hand a movement for more humane treatment. The public discussion of what we really want has barely begun.

In the secondary school

Most students of the secondary school are very interested in knowing more about biotechnology. They are confronted with aspects of it in newspapers, books, films, etc, but often they have a very undifferentiated view — either the very optimistic view of the technology as a solution to any problem, or the horror view of books like "Brave New World".

This is due partly to the lack of biochemical and genetic knowledge and partly to a very limited understanding of the technological development.

Secondary school Biology teachers in Denmark have gathered some experiences during recent years. Biotechnology concerns production as well as reproduction, each providing possibilities and approaches which should be part of education. Not only because the two topics are interlinked, but also because girls usually a priori are more interested in the productive aspects. The motivation for "the other aspect" emerges through a better understanding of the problems. Understanding biotechnology is not just a question of learning the underlying techniques and acquiring scientific knowledge. It is also necessary to understand the role of the technology in the society, the moral and ethical questions and which consequences the application of this technology might have for the environment, the individual, the country and the world community. A high level of understanding is reached only through this very holistic point of view. Futhermore an experimental approach is suitable to ensure that discussions do not become too theoretical and hypothetical drawing them away from the real problems to an irresponsible fantasy world.

A practical example

The following project is chosen to show how it is possible to work with practical biotechnological problems in the secondary school. The class had a great majority of boys, and the topic was genetic engineering in production. The purpose of the project was to enable the students to distinguish facts from fiction in information from radio, television, newspapers and popular magazines.

The students were asked to imagine themselves as a newly established research group in a factory producing enzymes for industrial use. They had a bacterial strain that — by chance — was found to produce a certain enzyme in large amounts. Having this strain meant that the enzyme could become commercially feasible to the factory.

The research group (*alias* the class) was now asked to answer the following questions:

— could commercial production of the enzyme be established?
— which kind of purification processes would be suitable?
— could the genetic information be characterised and integrated in a more convenient strain?
— could the factory take out a patent for the genetic information and the production?

(It must be noted that the strain was already well characterised by professional bacteriologists and recommended for this project).

A convenient assay for the enzyme and procedure for gel making, protein separation, and DNA extraction were found in the literature before starting the practical work. Preliminary experiments were performed to train the students in the necessary biological methods (cultivating bacteria, practising sterile technique, recognising the phenotype of the strain, determining enzyme activity).

The next step was to cultivate the strain in larger quantities for enzyme preparations. The students were organized in five teams, each working with different procedures of extraction. The methods were compared and evaluated in terms of yield and purity. The best method was used to prepare enzymes for further purification, which was done by precipitation. The product was not accepted by the class as sufficiently good for industrial use. The first two questions had been answered. The genetic information of the strain could not be transferred to other strain *in vivo*, but DNA-preparations could be made and analysed by gel-electrophoresis. A characteristic band of small-molecular DNA showed up on the gels: a plasmid. A new strain was then prepared for *in vitro* transformation and mixed with the DNA-preparations.

The result was overwhelming — hundreds of colonies with the expected phenotype. The genetic information (located on the plasmid) was established in the new strain, and the students were able to show that the plasmid had been assimilated by bacteria now having a high enzyme production. Thus the genetic information for high enzyme production had to be located on the plasmid, which answered the third question. The fourth and last question was only discussed theoretically due to lack of time.

Besides learning a lot about general biological phenomena and methods (in genetics, biochemistry and biotechnology), the students had to learn — and practise — laboratory safety rules: personal safety, responsibility to other persons entering the laboratory, and waste handling. Through the practical work they also learned to make critical evaluations of the information they got from different sources — and to argue intelligently about possibilities and problems of biotechnology in industrial production.

The reproductive aspects were not integrated in this project, but the class had previously studied sexology and human physiology and thus had a basis for spontaneous discussions of the relevant reproductive aspects.

Conclusions

The two examples from technology teaching in biology are both taken from low level biology teaching — each taking up one third of available biology lessons (both lasting about 12 weeks). They are rather large projects, but they incorporate aspects essential to technology education at the secondary school in my opinion.

Both examples included large amounts of practical work. The questions posed by the students were consequences of this practical work and therefore realistic and relevant to the actual problems in the use of the technologies discussed. They also fulfil the theoretical considerations presented in Fig. 1 — to view technology as part of the activities in society — not only affecting production, but also the workers, the population, the environment, and the economy of the society.

Knowledge of a particular production or a factory can also be achieved by visiting an establishment. But it is my experience from short visits that these often tend to be too superficial for secondary school students, and that longer visits — 5 to 10 days — are only accomplished with many obstacles. Only a few factories are interested enough to take the trouble, and the school system of Denmark is not adjusted to this kind of education.

In many cases, I find it more convenient to make the actual production in the school laboratory on a small scale. It is easier for the students to survey the different parts of the production and the technology used, and it is possible to take a break for a while in order to solve theoretical problems at the appropriate moment, not several days later, when some of the impressions have vanished. A combination of a "home-production" and a visit provides good knowledge and understanding of applied technology. Understanding the role of technology in the development of the society requires knowledge from other subjects than the sciences (economics, history and geography), whereas evaluating technology requires ecological, sociological, political and ethical considerations. Limitations in school structures impede a total integration of different subjects. In practice it is therefore necessary to work within a less idealistic framework as demonstrated in the cases above.

These considerations concern both "traditional" technology and the "new technologies". But whereas the consequences of traditional technologies are well described, very little is known about how "new technologies" will influence employment, pollution and the structure of world economics. At the present time, biotechnology is only developed in the richest countries of the world, but could it possibly be a technology for the developing countries in order to speed up economic growth? Or will it on the contrary deepen the cleft between rich and poor countries in the view of the existing world economic order? Can biotechnology be used to solve the problems of malnutrition, or will biotechnology production replace the most polluting productions in the rich countries, while these are transferred to developing countries causing environmental problems there? These unanswered questions might seem speculative at this moment, but it is the students of today who are going to make the decisions.

Postscript

While writing this article, the first application for permission to start a commercial production of a human protein (insulin) from bacteria was submitted in Denmark. The bacteria are a result of genetic engineering. It is unknown when the permission will be given, because the county council did not find itself competent to treat the application!

Acknowledgement

This paper was read at the Nordic Conference on Science Education, held in Copenhagen, May 1985, and is reproduced by kind permission of the author and the organiser of the conference, Professor Erik Thulstrup of the Royal Danish School of Educational Studies, Copenhagen, Denmark.

14

Technology Education: A Union of Science and Technical Skills

J. B. HOLBROOK
University of Hong Kong

Clearly there is a case for interlinking aspects of both science and technical studies into a single course. This would place a greater emphasis on producing a final product than would a traditional science course, and it would include more "finding out" experimentation than a traditional technical studies course. The course could be described as one applying science to produce something useful, a description very close to a definition of technology as the process of producing something useful through the application of knowledge and skill. Such a course would have an intellectual dimension and a practical dimension, which would involve process skills.

Whereas many science courses provide an academic background for further studies, this technology course would be geared more to the needs of the society in which the course is taught. Technology is not technical studies and, whereas technical studies demand workshop facilities, this course would operate on a smaller scale and diversify to include a wide range of manipulative skills.

The objective of such a course would be similar to those put forward by the Government of Pakistan for Technology Education[1]. They are:

1. to prepare students for making useful contributions to home, school and community life,
2. to encourage students to explore and experiment and to express individual creativity,
3. to develop in students systematic, clean and safe work habits,
4. to create in students an awareness of their aptitudes, their abilities and their inclinations,
5. to develop in students the habit of engaging in gainful pursuits for making the best use of their time,
6. to develop in students an appreciation of the value of private and public property, and of their preservation,

7. to impart to students a knowledge of basic materials, processes and techniques in terms of their usefulness in everyday life,

8. to develop in students a proficiency in the use of elementary hand-tools and equipment,

9. to inculcate in students an appreciation for the cost of materials in order to reduce unnecessary wastage,

10. to develop in students an appreciation of the role of agriculture and industry in the development of the country,

11. to impart to students a knowledge of the needs of conservation in the utilisation and development of natural resources in a country.

However, more traditional science education objectives are also suggested for such a course.[2] These are:

1. to increase the student's knowledge of basic concepts in science, the physical sciences, life sciences and earth sciences,

2. to develop facility in using the methods and tools of science,

3. to promote an understanding of the role that science plays in the development of societies, and the impact which society has on science,

4. to develop student appreciation of science as a way of learning and communicating about oneself, the environment and the universe,

5. to assist a student to develop as an autonomous and creative individual who will live in a scientific and technological society,

6. to create an enthusiasm for the method of thinking which uses observed facts in the logical solution of problems,

7. to expose students to a representative sample of technological applications of science — communications, transportation, scientific research, medicine, architecture, computers, household appliances, energy,

8. to relate science to career opportunities in technology, industry, commerce, business, medicine, engineering, education, research and other areas in which science plays a role.

Throughout the proposed course an emphasis is placed on producing something in its final form rather than simply illustrating a process. This is not an easy task: one may have to spend a considerable amount of time acquiring a manipulative skill in order to obtain a product. Three suggestions follow, which might be appropriate at junior secondary school level.

Materials from plants

Under this topic, four products are considered:

(i) soap,
(ii) perfume,
(iii) plant dye,
(iv) paper.

These products are brought together in order to end up with a wrapped, coloured, perfumed bar of soap similar to that available in the local shop. It is hoped that the study would achieve the following:

(a) appreciation that a bar of soap can be produced by:
— the simple process of reacting plant oil with an alkali,
— colouring the soap with a plant dye,
— perfuming the soap with a perfume extracted from plants,
— forming a bar of soap in a mould (through construction of the mould),
— wrapping the soap in paper (by producing paper from plant materials),
— glueing the paper around the soap to make the finished product, (extensions are possible, for example silk screen printing on the paper).
(b) awareness of the history of soap production and tools of science,
(c) understanding of various conceptual factors in the manufacture of soap, perfume, dyes, and paper, and awareness of what will happen if certain factors are varied,
(d) development of construction skills by making a soap mould and a mesh framework for the production of the paper,
(e) development of manipulative skills through handling the chemicals needed to make soap and for testing the products,
(f) awareness of economic, social and ethical aspects in the production of home-made and industrial soap,
(g) understanding of the cleaning action of soap under various conditions and in comparison with other detergents.

The technical skills involved will be in the construction of the mould (it can be made from wood or plastic) and a framework for paper making (a wooden frame with suitable joints, covered with a wire mesh).

The scientific skills involved would include the making of samples of soap (possibly from different starting materials); simple steam distillation to extract perfume (for example, from orange peel or flowers); solvent extraction of plant dye; breaking down plant fibres and dissolving non-

EIT-D*

fibrous material to produce paper; handling sodium hydroxide or producing an alkali from wood ash.

Fibres and fabrics

Under this topic, the following are considered:

(i)   making a thread by spinning,
(ii) weaving thread to produce cloth,
(iii) dyeing the cloth,
(iv) production of regenerated and man-made fibres,
(v)   recognition of fibres and fabrics.

It is hoped that the study would achieve the following:

(a) appreciation how plant material can be made into cloth,
(b) awareness of various forms of dyeing,
(c) appreciation of the need for and the production of regenerated and man-made fibres,
(d) development of a "feel" for fabrics,
(e) understanding the relative merits of natural and hand-made fibres, and the reasons for mixing fibres,
(f) the ability to assess the suitability of cloths for a variety of conditions (for example, moisture retention, resistance to wear, crease resistance, strength of cloth).

The technical skills involved will include testing of strengths of fibres using a simple machine (a problem-solving activity), testing the fastness of dyes (making a simple washing machine), tie and batik dyeing, testing the wear resistance of cloth (involving the construction of a test rig), constructing of a weaving frame and the process of weaving.

The scientific manipulative skills involved would include the examination of fibres under a microscope, production of regenerated and man-made fibres, dyeing cloth, testing fabrics.

Cosmetics

This topic is really an alternative title for the teaching of emulsions and colloids. The following are produced as finished products:

(i)   two emulsions (hair oil, handcream),
(ii) one colloid (toothpaste),
(iii) shampoo and lipstick for classification as either colloid or emulsion.

The goals for studying this topic are to understand and recognise emulsions and colloids; to appreciate the reasons for using them; to appreciate the economic, social and ethical considerations surrounding the use of cosmetics and the need for safety standards; to understand the need for care of skin, hair and teeth.

The technical skills involve the production of moulds for making lipstick, etc; packaging cosmetics; silk screen printing (the importance of appearance).

The scientific manipulative skill involved would include making emulsions (oil-in-water and water-in-oil), making colloids, tests for emulsions and colloids.

References

1. Unesco, Technology Education as part of General Education, A Study based on a survey in 37 countries. Science and Technology Education Document Series, No. 4, Paris, 1983.
2. G. W. F. Orpwood and Jean-Pascal Souque, Science Education in Canadian Schools, Background Study 52, Volume 1, Science Society of Canada, 1984.

15

The Place of Small-Scale Industry and Technology in the Curriculum of Junior Secondary Schools in Developing Africa: The Case of Nigeria

O. C. NWANA

University of Nigeria, Nsukka, Nigeria

Secondary schools in Nigeria now offer both literary and technical subjects to all students. Junior secondary schools provide 3-year courses in which pre-vocational subjects are taught together with regular subjects. In senior secondary schools the 3-year courses provide some specialisation either in academic-type or in vocational subjects.

The new national policy fashioned at a time in which Nigeria had a favourable economic climate had a vision of laboratories and workshops across the country where the students could learn trades. Now with a faltering economy, schools still do not have laboratories, workshops or technically-trained teachers.

For the new policy to become effective, is it possible for such training to be done within the community? School and Society have over the years tended to be separate entities, often one being ignorant or unmindful of what the other is doing or is about. Indeed the School has often been screened from Society by its four walls, which have progressively got thicker and higher. The proposal is to break through these walls and let both School and Society interact to solve mutual problems. In recent years the government has shown that it cannot continue to employ school leavers (the jobs are just not there) and more importantly that it cannot retain the number of civil and public servants under its employment on the salaries fixed during the years of plenty. Galloping inflation has progressively reduced the purchasing power of salaries paid to teachers, clerks, administrators, nurses, workers, etc. The money they earn quickly passes into the hands of tradesmen with much less, if any, academic qualifications.

A list of small-scale industrial/technical endeavours which characterise the urban areas of the country will throw light on the picture:

1. Auto-mechanics
2. Domestic electric wiring
3. Radio/TV repairs
4. Cloth-dyeing
5. Wood carving
6. Pottery
7. Weaving
8. Plumbing
9. Brick-laying
10. Vulcanising
11. Bakery
12. Tailoring
13. Carpentry
14. Iron-welding
15. Building drawing
16. Brick-making
17. Leather tanning
18. Shoe-making
19. Interior decoration
20. Printing

An increasing number of young people are making much more money in these areas than do salaried public servants. Besides, they pay little or no tax as their total earnings are often difficult to establish.

Handwork/Handcraft were previously discredited aspects of the primary school because they did not pay. Technical subjects in the secondary schools were left to weak students who subsequently obtained lower paying jobs in the public service. But the situation today is different. Public sector jobs are on the decline while private sector jobs are on the increase. People are being called upon to demonstrate self-reliance (a concept which President Julius Nyerere of Tanzania has promoted for three decades) and the educational system is being called upon to demonstrate relevance as the old skills are not useful any more to those who possess them. The call has therefore been made to build into the curriculum instruction in the small-scale industrial skills of the tradesmen within each community with the master tradesmen or master craftsmen/women as the principal instructors.

Implementation

The plan is as follows:

(a) Locating and listing of all small-scale industries in any area.
(b) Evaluation of each industry in terms of how many apprentices it can take on effectively without disturbing the normal flow of business.
(c) Estimating how much money each industry makes in the month.
(d) Assessing how much extra equipment can be installed in the industry to widen its scope in giving worthwhile experiences to pupils.
(e) Re-scheduling the school timetable to make available long blocks of time (flexible time scheduling) for practical training.

(f) Scheduling regular meetings of tradespeople with officials of the Ministry of Education and the Local Government Authority to ensure smooth running of the programme.

Only a few school communities have been able to initiate action in this plan. Some of the major problems so far encountered are as follows.

1. Even though some tradesmen are thrilled by the idea of teaching school children, i.e. of the school moving to their workshops, they are not ready to adjust their teaching strategies to those normally carried out in the schools. For example, the master auto-mechanic simply carries on with his activities, requesting one spanner or tool from the apprentices as required by the job and hardly ever explaining a particular technique or approach. He expects the apprentices to be observant and draw their own inferences. There is no dialogue, certainly no questions are tolerated as one would expect in a typical science/technical laboratory. The regular apprentices are used to this system, but the visiting school apprentices are not content with this conservative mode and their rather frequent questioning does lead to conflict situations between them and the instructing tradesmen.

2. Most small-scale industrial firms are in small sheds and pre-fabricated enclosures not suitable for large-scale instruction. The tradesman have requested allocation of land as well as bank-loans to enable them to expand their establishments for coping with their new instructional roles. This has not been forthcoming due to limited financial resources available to Government and the fact that in some communities it is impossible to release land for such use.

3. Some of the tradesmen are sceptical of the project in the sense that they are afraid that government officials will thereby know about their business and perhaps thereafter levy heavy taxes on them (a real possibility).

4. School children have been known to demonstrate less than ideal behaviour when assigned to the industrial firms for they are not under the same strict disciplinary control as would be the case under normal classroom conditions. Indeed the number of pupils attached to each trade is such that many of them mill around doing nothing, if not obstructing the progress of others. There are simply too many for effective control.

5. The normal school scheduling of lessons on a 30-minute or 40-minute basis is most unsuitable for acquiring these trade skills. Thus pupils attached to bricklayers need to be there for long periods such as two-week periods at a time (full-time) if anything can be learned. This conflicts with other subjects offered in the school. There is need therefore for more thought to be given to the concept of differential scheduling for the various trades.

6. There is a growing need to bring the master tradesmen to occasional in-service courses in which they can be introduced to the fundamentals of formal instruction as a supplement to their informal mode. This becomes necessary as the tradesmen usually demand compliance from the students; they use abuse and physical assault as means of disciplining apprentices rather than more psychologically acceptable methods. But there is the danger that too much formality may destroy the talent in these well-tested and competent tradesmen.

7. Many parents think that present economic depression is temporary and that soon it will be over and that school leavers can once again aspire to well-paid official jobs. But this is far from the truth. The indications are that harder times are to be expected. Where parents think that industrial-technological training is a short-term stop-gap, there is a need to embark upon massive adult-education in which their attitudes are appropriately modified or more properly aligned. The task of public enlightenment is yet to commence in Nigeria and most of developing Africa.

Making science teaching cash-productive

Allied to giving school children industrial and technological training within existing small-scale industries is a project which the Science Teachers Association of Nigeria is sponsoring under the name "Making Science Teaching Cash-Productive". In this project, the Science Teachers Association of Nigeria has begun a campaign for the modification of the curriculum of the regular science subjects in such a manner as to make their teaching productive rather than being constantly a source of consumption, wastage and economic drain. The project was necessitated by the fact that the science departments in the schools are no longer given enough money for equipment. Money-yielding projects therefore come in handy in enabling the schools to sustain science teaching. Examples of the activities are:

1. The physics department would establish commercial battery charging units and undertake electrical wiring of houses.
2. The chemistry department would produce and bottle distilled water, battery electrolyte, make soaps and detergents, re-charge fire extinguishers, make insecticide.
3. The biology department would cultivate ornamental plants for sale, breed pets, maintain a zoological garden.
4. The agricultural science department would run a poultry farm, operate commercial grains and root-tuber farms and maintain some livestock, make livestock feeds, etc.

In this project practice rather than theory is the major determinant of curriculum content. The science teacher would teach more of those topics which have practical applications relevant to contemporary society than topics of purely theoretical value. This approach is dictated by adverse times and if properly carried out would provide the much needed bridge between science and technology, the one field enriching the other and benefiting therefrom.

There is a great future for operating industrial/technological training within the science curriculum of primary and secondary schools in developing Africa. The enrolments in scientific and applied-scientific disciplines in tertiary educational institutions is disappointedly low. If the project succeeds, science teaching would become revitalised, pupil interest greatly heightened, while industry and technology would become truly a way of life of the people rather than an occasional enterprise for only a few as has hitherto tended to be the case. This would be one way of achieving the goals of science education in providing for the future needs of humanity.

16

Physics Plus: One Way of Linking School Physics with Technology

F. R. McKIM

Marlborough College, UK

As described many times in this book, Physics courses for students up to 16 years of age tend to be rather pure, concentrating on the teaching of basic concepts. This teaching is largely by theoretical discussions, supported by laboratory experimenting. There is not much emphasis on the way in which these concepts are applied in the world at large.

Analysis of questions set in national examinations in physics in the UK for students at 16+ (Grade 10) suggest that applications are not a significant feature and little understanding of applications is expected of students or needs to be demonstrated by them at that stage. Various bodies have expressed concern at this. But no easy means of changing the state of affairs has as yet been proposed. One proposal has been to replace courses in physics by courses in technology. While this proposal has had enthusiastic advocates it has not gained wide popularity.

One reason for this seems to be a feeling by physics teachers that they are not adequately qualified to teach a complete course in technology. Another is that if a school lays on a course in technology it usually feels that it needs also to continue to provide a course in physics; this means that students have to choose at an early age which course they are to take.

The Physics Plus project[1] is attempting to solve the problem in a different way. We are producing material written specifically for students, on a wide variety of topics, each of which represents a situation involving some applications of basic principles in physics.

In the past it has been the usual practice, if applications were treated at all, to have a few pages in each chapter of the school textbook devoted to particular topics. But there is a fundamental objection to this practice. Applications of physics inevitably go out of date quickly, whereas fundamental principles do not. If the applications which are discussed are to be kept up to date, then the whole textbook must be thrown away every 3 or 4 years and a new textbook, with up-to-date applications introduced. This can be prohibitively expensive.

However, if applications are printed in the form of pamphlets, then if a particular application goes out of date it is comparatively cheap to replace the pamphlet with a new one. (And the school textbook which deals just with the fundamental concepts can remain in use).

The story in each topic is told within a total length of four pages in pamphlet form, so that even the reluctant student should not find the task of reading it too daunting. Each pamphlet is being written by a school teacher together with a technical expert from the field being written about. The intention is to produce about 100 pamphlets.

One can get some idea of the wide range of topics from this list of titles already published:

Karate	About sailing
Modern buildings	The diesel engine
Timetabling express trains	Energy from the tides
The efficiency of large trucks	Electric light bulbs
Physics and sport	Electrostatics outside the lab.

Two examples can be given to indicate the kind of treatment of a topic within a pamphlet.

"Timetabling Express Trains" considers the problem of introducing High Speed Trains on to a track which already handles slower trains, and where there are no loops nor many junctions. The problem could of course be solved algebraically, but the treatment is in terms of distance-time graphs. The discussion is based on the line from London (Euston) to the NW of Britain, but no familiarity with this is essential. Questions are asked on redesigning the timetable for this route.

"Electrostatics Outside the Lab." deals mainly with (i) the removal of dust from power station exhaust gases by electrostatic precipitation and (ii) the precautions taken to make sure that there are no explosions on board crude oil super-tankers during tank cleaning operations. There are passing references to bodies becoming charged by friction against artificial fibres (for example in carpets or car seats). In each case explanations are linked to the basic concepts likely to have been treated in school courses.

Other topics among the total of 40 so far published (1987) include

Alternatives to petrol	Beyond the bunsen burner
The Thames Barrier	Greenhouses
Railway Lines	The bicycle
Light and the motor car	Heat exchangers
Seeing without being seen	Producing TV pictures
The Hay inclined plane	Mining hydraulics
Fire Alarms	Using radioactive tracers
Applications of Ultrasonics	Predicting volcanic eruptions

Convection around your body   Chimney changes
Distribution of gas   Rock climbing

As will be seen from the examples given, some of the topics (e.g. "The Thames Barrier" and "Energy from the Tides", which deals with a possible tidal barrier in the Severn Estuary) are specific to the UK and would mean little to students from other countries. However, similar examples are possible for any country in the world where physics is taught. Certain topics, such as stability when engaging in karate or another of the martial arts, are the same the world over and immediately appreciated by anybody. Other technological applications of physical principles will be more specific to particular regions. What is needed is the right attitude of mind on the part of teachers to look for these applications and then to write them up.

The educational intention behind this whole enterprise is to dispel the idea that at a school level students should see physics (or science) as one subject and technology as another. This is a false antithesis. The idea is better described as a spectrum. Pure science is at one end; there is then a gradual change, but never a break, between that science and its applications in a wide variety of fields. The hope is that pupils in the future will be able more clearly to see that activities in technological fields depend on, and clearly link with, the basic concepts of science which they study in their school courses.

They need both technology and science, and I suggest that developments like the Physics Plus project will help them to get both.

It is proposed to set up an organisation which will continually review the Physics Plus pamphlet topics, and produce new ones as some of the existing topics go out of date.

Reference

1. Details of Physics Plus material can be obtained from the Project Director, F. R. McKim, Marlborough College, Wiltshire SN8 1PA, UK.

17

Extracurricular Activities: Some Developments in China

SUN RUOHAN

The 16th Middle School, Tianjin, China

Science and technology are developing by leaps and bounds all over the world, posing a new challenge to scientists and educational workers. Our government has all along looked upon bringing up young people as a reserve force for science and technology as one of the important measures for development. Education is an enterprise for the future, and in running schools not only should we have the present in mind, but also the future. For this reason, we should devote major efforts to the development of extracurricular activities of science and technology, developing intelligence and fostering a new generation of gifted people with creative ability who are readily receptive to new ideas, new concepts and new science and technology, and adapted to various social changes and reforms. We are actively developing extracurricular activities of science and technology, as described below, for these reasons.

In 1981, the National Youth and Children's Scientific and Technological Activities group was set up, consisting of leadership from the Chinese Science and Technology Association, the Ministry of Education, the Central Communist Youth League, the State Physical Culture and Sports Commission, and the All China's Women Federation. At the same time the National Youth and Children's Science and Technology Counsellor Association was set up, followed by the corresponding organisations established in provinces and cities, as well as in schools, thus pushing forward and guiding the scientific and technological activities of young people and children, imparting to them a rudimentary education in science and technology, fostering scientific quality, and training an enormous reserve force for science and technology for the modern construction of the country.

The successes achieved by our school, which were derived from the scientific and technological activities, were under the Group Leadership of the Tianjin Youth and Children's Activities. At a conference on "Science for All", Unesco commended our work, stating: "These delivery systems can be and are used in an extracurricular way by the formal system. The extensive extracurricular programmes being utilised in China might serve as a model for other countries in the region."[1]

The activities are based on topics which occur in industrial and agricultural production. They aim to encourage a thoughtful approach to knowledge, as well as fostering good habits of learning, a love of science and various creative abilities and skills, through observation, practical experience and discussion by the students under guidance from teachers. The activities depend on voluntary participation by students, grouped according to interest and hobbies. They may involve activity in the natural environment, in local factories, in laboratories or in a society at large.

First of all, we launched a popularisation campaign, using science broadsheets, wallcharts, show-windows and small-sized newspapers of science and technology run by students and teachers. In order to popularise scientific and technological activities, we have fixed October every year as "Love Science Month", in which each student is required to do "Six Items", namely to study a new scientific phenomena; to get to know a story about a scientist; to read a technical book as background reading; to observe and describe a natural phenomenon; to make a scientific and technological product; to conduct a small scientific and technological experiment; and to write a small scientific treatise or composition.

In activities in "Love Science Month", we have held scientific and technological exhibits, which have greatly stimulated students' interests in exploring science: they can draw a majority of students into participation. We have organised various interest and subject groups. The eight interest groups in our school which have drawn the participation of students are radio, model airplanes, photography, solar energy, meteorology, growing, planting, scientific products. There are twenty-one subject groups and they are set up according to the grades of the students: the subjects include computing, mathematics, physics, chemistry, biology and geography. Some students do exceptionally well in these activities, thereby consolidating and deepening the knowledge learned in class.

To stir up students' enthusiasm for learning, besides some activities organised inside the school, we have also arranged to take part in local and national competitions and in activities at summer and winter camps arranged by our city. A schedule for extracurricular scientific and technological activities is specially prepared by the school so that no student is allowed to join more than two group activities, so that adequate time is ensured for each. The students in our school now joining extracurricular science and technology activities include 83% of the total number, thus making the extracurricular life rich and colourful.

Statistics on participation in the Six Items

ITEMS / GRADE	Total number	Activities organised by the School						Students' individual activities						Books read	People fulfilling the Six Items	
		Subject and team meetings		Scientific films and slides		Lectures on special topics		Scientific products		Small experiments		Scientific compositions				
		No. of times	Partici-pants	No. of times	Partici-pants	No. of times	Partici-pants	Small prods	Small inven-tions	Items	Partici-pants	Compo-sitions	Papers		No. of people	Percent-age
Junior 2nd grade	219	3	659	1	219	14	560	213	16	3	60	219	4	535	219	100
Senior 2nd grade	245	3	735	3	735	33	2098	205	17	2	240	245	41	614	224	90
Whole school	1415	12	2743	11	2661	171	12395	1282	93	23	1120	1505	120	3016	1278	90

Successful results

1. Raising the quality of classroom study by the students

For example, in the national enrolment examination of colleges and universities in 1982, 92.5% of our graduates went to schools of a higher grade. All the 44 students of the maths group entered the maths departments of universities. Of them 42 went into the maths departments of key universities. In the national maths competition of middle school students held in 1983, our students came first and second in the local maths competition. In the local physics competition of the middle school students held in 1983, it was also our students who came top of the list, and all the 32 students taking part in the competition were awarded prizes, making up 34% of the total prize-winners.

2. Training a reserve force for science and technology

Extracurricular activities of science and technology have broadened students' horizons, fostering their creative power and spirit for studying assiduously, thus laying a foundation for engaging in the cause of science in the years ahead. Many of them have been trained to have their own independent views and to be bold enough to overcome difficulties. Some students from our school, Wang Yuan-loong, Xu Hong and Xu Mei, have come first in National Prize competitions.

3. Raising the abilities of teachers

Teachers from our school are currently the scientific and technology activities counsellors in our school. In 1981 we invited 32 teachers to become counsellors and by 1984 the number had increased to 73, accounting for slightly more than 50% of the total number of the teachers in the school. They have a strong sense of responsibility, they work hard and are not upset by criticism. Some of them have participated in study classes and some are studying on their own, thereby keeping up to date in scientific developments and methods of teaching.

Basic requirements for organising extracurricular activities

The leadership of the school must clarify their ideas about running the school and be clear about the importance and urgency of these extracurricular activities. The activities should be orientated towards what is modern, towards the world and towards the future.[2] The contents should be rich in appeal so as to develop students' enthusiasm. It is necessary to mobilise the whole staff and workers of the school to bring the activities into line with the working plan of their departments and establish a regular supply of counsellors, and to assist actively in creating and developing appropriate conditions for scientific and technological activities.

References

1. *Science for All — Report of a Regional Meeting.* Unesco, Bangkok, p. 25, 1983
2. *Three Orientations for Running Jing-shan Middle School,* proposed by our Leader Deng Xiao-ping. Beijing, 1983.

18
Vocational Science Curriculum Development in Thailand

MANEE CHANDAVIMOL

IPST, Bangkok, Thailand

The Institute for the Promotion of Teaching Science and Technology (IPST) in Thailand has developed science curricula in schools at all levels since 1971. From 1971–1978 projects undertaken were the development of mathematics for primary and secondary schools, and six projects on science in lower and upper secondary schools. In 1979 IPST was asked to develop vocational science and mathematics curricula for the upper secondary level. This paper will discuss particularly curriculum development in the sciences.

Vocational education is divided into two levels, an upper secondary level course of 3 years' duration leading to a certificate in vocational ecucation and graduates are qualified to work as craftsmen. A further 2 years at college level leads to a diploma in higher vocational education, and graduates are qualified to work as technicians. Further vocational education for the Bachelors' and Masters' degrees is obtainable from colleges and universities.

The need for the development of curricula in science and mathematics in vocational schools arises because the existing curricula do not relate to vocational topics. In addition, there is a lack of textbooks, equipment, and appropriate methods of science teaching and learning.

This activity was started in 1979 at IPST through the analysis of the necessary science and mathematic concepts and principles at the upper secondary level for industrial arts, agriculture, commerce, home economics, and arts and crafts. Vocational educators, science supervisors and teachers, and instructors from the tertiary level together with IPST curriculum developers participated in an IPST workshop to define science and mathematics concepts for vocational areas. National and international consultants also participated in these discussions. Using the resulting analysis and suggestions, objectives were agreed and criteria identified for the selection of course content in each vocational area.

Objectives and course design

The main objective of science courses for vocational training is to develop an understanding of science concepts related to each vocational area. At the same time, science process skills and scientific attitudes are also developed so that the students will be able to solve problems in their vocations as well as in everyday life[1]. The nature of the science courses for vocational students will differ in depth from those developed for the academic stream; only basic theories and principles are required, with applications relevant to the vocations. The disciplines are integrated as much as possible, and the courses are mainly experimental.

The methodology of teaching is planned so that the teachers will teach science through an inquiry approach. The emphasis on the inductive reasoning process is decreased, while basic knowledge and deductive thinking are stressed. In other words, the theoretical approach is minimised and problem-solving is emphasised. The vocational science courses were designed within the framework of the objectives serving the students' vocational interests. Content and activities are integrated all through the course. Experiments and other activities aimed at illustrating the basic concepts in science which are important to the vocations were included. Equipment prototypes were designed locally, and some of the inexpensive equipment developed for the academic stream by the IPST also served the courses.

Industrial arts students are required to take four semester science courses. Since there are six major trades — mechanics, welding, metalwork, electronics, electricity, and construction — all students take a common core course which mainly covers mechanics. In the following semesters, the necessary content becomes different; consequently, separate courses are required.

Agricultural students are required to take four semester science courses: two in biology, one in chemistry and one in physics. Home economics students are required to take four semester science courses, whereas arts and crafts students take three. The core courses were designed to cover basic concepts and principles in physics and chemistry as well as supplying the necessary scientific information for these two vocations. Home economics students need two additional courses in food science and basic concepts in nutrition. All courses are in modular form. The commerce curriculum has also adopted the four modules of the IPST physical and biological science courses as they contain up-to-date scientific knowledge for daily use, including practice in science process skills.

The design team

The vocational science curriculum design teams consisted of a representative cross-section of vocational science teachers and supervisors, instructors from institutes of technology and vocational education, college instructors and university lecturers. Training programmes for the design teams took a variety of forms, including on-the-job-training, workshops and seminars, and Unesco study-tour grants for the key personnel in each team.

Development of materials

Trial student textbooks were drafted with integrated content including experiments and activities. The writing style emphasised a problem-oriented approach as much as possible, quite different from the traditional texts. Examples and applications were related to vocational subjects. Teachers' guides were also written concurrently with the second draft of the text, and the developed materials were tried out in vocational schools and colleges during the 1979–1980. The trial version of the teaching materials was then prepared by each design team on the basis of the experience obtained in the trials.

The key to the success of any curriculum development is the teacher. Therefore, an in-service teacher training programme was organised by IPST to familiarise teachers with the materials developed, the new teaching approach, and new laboratory equipment. The in-service training programme adopted a teaching approach which integrated content with methodology. Teachers had the opportunity to perform all the experiments which they would have to manage in the classroom. Techniques of teaching and evaluation were provided, together with visits to certain industries and places of interest to enhance their experience in each vocational area. Instructors attempted to use the same methods that they would like to see teachers use in their own classes.

Follow-up programme

Nationwide implementation of science curricula in vocational schools began in May 1981. The IPST scheduled a follow-up plan by sending a group of design teams to visit randomly selected schools throughout the country. These tackled teachers' classroom problems, held teachers' meetings, and discussed any problems concerned. Students were also randomly requested to complete questionnaires for each chapter of the texts used. This feedback information was analysed and will be used in the revision. Summer workshops for teachers who taught the new science curriculum were arranged at IPST with the objective of obtaining first-hand information about teaching and learning situations in the classroom.

Reference

1. Nida Sapianchai, Science Education in Secondary Schools in Thailand, A Position Paper, The Institute for the Promotion of Teaching Science and Technology, Thailand, 1983.

19

Women in Science and Technology Education

I. J. MOTTIER and J. H. RAAT

Ministry of Education, Leiden and the University of Technology, Eindhoven, The Netherlands

In most countries girls and women participate far less in science and technology education than boys and men. This is the situation in The Netherlands and several projects have been set up there in order to promote girls' participation.

Because physics is often taught in such a way that it conveys a male image (the teachers are male, the applications and examples of physics are chosen from male interests), the MENT project developed special curricula more attractive to girls by paying more attention to things of daily life, of the human body and of social implications of science. The project also provides in-service training for teachers in order to make them more aware of the situation over girls in physics and technology.

There are barriers which discourage girls from studying science and technology: they sometimes give up mathematics or physics at too early a stage and this makes subsequent technological studies impossible. Another barrier may be caused by teachers who make girls themselves believe they have no capacities for science. Another barrier may be found in the absence of good counselling. Yet another is the lack of suitable textbooks.

Projects have been established in The Netherlands to reduce these barriers, for example the Handrover project which is producing guidelines for sex equality aspects in textbooks and in curricula, at the same time promoting awareness of the social implications of science and technology.

Another initiative in The Netherlands has been the establishment of courses to promote women's participation in traditionally male professions: electricians, furniture-makers, plumbers, fitters and painters.

The rapid development of information technology is leading to its introduction into education. Women can participate equally in this development and this is being given attention in The Netherlands. A support group on "Women and informatics" exists, as well as other support groups such as "Women and mathematics", "Women and science" and "Women in the technical professions".

Industrial and Technological Issues: Examples of some Secondary Science Curricula

Introduction

The rationale for the inclusion of industrial issues in science curricula and the strategy by which they may be taught, may be common in different parts of the world, but the approach and contents must depend on the particular local context. In this section, there are examples from Austria, Israel, New Zealand, Papua New Guinea and the Philippines. The example from the United States, the Chemcom course, is different in that it is written for the non-science specialist. Towse describes how resource materials can be made available to students, using his experience in Zimbabwe. Finally, Yakabu describes some of the progress made in a new course he is developing, Science in Ghanaian Society.

20

The Development of a Relevant Chemistry Curriculum for Austrian Secondary Schools

E. M. JARISCH
Bundesministerium fuer Unterricht und Kunst, Vienna, Austria

Late in 1983, Project Group Physics/Chemistry (PGPh/Ch) began work developing curricula for physics and chemistry for the basic level of secondary schools. The purpose of this was to co-ordinate the presentation of topics in physics and chemistry in order to facilitate the transfer of students between the Hauptschule (HS) (aged 11–14) and the Allgemeinbildende Hoehere Schule (AHS) (basic level 11–14 and advance level 15–18). For this reason, the Project Group contained treachers from both HS and AHS as well as teacher trainers.

The goals of the PGPh/Ch were to identify ways and means to achieve the following:

(i) an understanding of the relationship between physics and chemistry and everyday life and problems associated with the working place;

(ii) an understanding of the relationships between ecological and economic problems as they relate to physics and chemistry.

Therefore, the critical task of the PGPh/Ch was the identification of those topics which could lead efficiently and effectively to acheiving these goals while satisfying a number of constraints. The major constraints to be considered were: first, the age group of the students is 11 to 14; second, the technologies in chemical processes are advancing continually; third, those students who complete their education with the polytechnic year (age 15), enter their working life without expectation for further chemical education; and finally, a strong tradition of humanistic goals in the Austrian educational system sets severe limits to the availability of teaching units for chemistry.

Given these goals and severe constraints the present author proposed a particular way of combining certain topics in chemistry with topics in physics. The main idea of combining topics is the increased efficiency of presenting a large body of knowledge. More specifically, the selection of combinations of topics has to concentrate on problem areas in which chemistry and physics are highly interwoven. In many instances they are in fact indistinguishable. For example, consider the topics "Structure of Atoms", "Electricity and Ions", "Conducting and Nonconducting Materials", "Combustion and Heat Energy", and "Properties of Materials".

In preparation for these topics the students have their first year of physics in the second year of the basic level. Here they learn qualitatively about different types of movements, about energy, inertia, and forces. In their third year of basic education they also learn about electricity. Finally, in their fourth year chemistry is independent of physics. Efficiency and effectiveness in this fourth year are accomplished through differentiation in terms of selected topics following the preferences of students and teachers, the opportunities of the region, and environmental concerns. The selection of topics, however, has to satisfy certain minimal requirements for the AHS and the HS. Generally, more freedom is given to teachers in the HS about concentrating on a limited number of topics. In view of this freedom one distinguishes between topics useful for "teaching-projects" and topics of which only a more limited knowledge is necessary.

Examples of topics with a high degree of flexibility regarding the detail of their presentation are: production of aluminium, iron, fertilisers, construction materials, coal, and petroleum. These topics also consider improving the utilisation of natural products and energy. In contrast to these topics, the concepts of certain fundamental reactions are regarded essential, for example, acid and base reactions, oxidation and reduction, or polymerisation, and little freedom is provided about choosing the extent of their presentation.

Modern technology and teaching chemistry

Curricula prior to 1962 required the presentation of specific technological processes, many dating back several decades. Teachers had no freedom to replace the discussion of outdated processes with newly developed process technology. The PGCh recognises this limitation of the traditional approach to curriculum development and began searching for new teaching concepts. The new approach to curriculum development rests on two ideas: first, the curriculum does not limit teaching to specific process technologies but gives recommendations to discuss relevant problems. Second, the PGCh developed close ties to the chemical industry in order to find support in identifying modern technological processes and engage it in supplying teaching materials.

Today, the Association of the Austrian Chemical Industry (Fachverband der Chemischen Industrie Oesterreichs).

Such support by the FChIO is highly appreciated in view of a recent poll (1982) by the Gallup and the Bartberg Institutes concerning the image of chemicals and chemical industry in Austria. The result showed a very inconsistent view of the need and use of chemicals versus their production. On the one hand a variety of chemicals are desired for every day life; on the other hand a strong resentment to the chemical industry exists. Psychological tests revealed that chemical processes and reactions are associated with responses like "black", "bad", "poisonous" and even "horrible".

The one-sided perception of chemistry shows the need for better education in chemistry. In particular students have to be made aware of the importance of chemistry and chemical reactions in many natural processes which are fundamental in biology. Given the limited success of past methods to achieve a fair public awareness of the significance of chemistry the need for new approaches to teaching chemistry are underlined.

Topics in the curriculum

The basic chemistry course in AHS and HS for students for the 3rd and 4th years of the basic level course are:

Year 3: *Properties of materials.* In nature we find only rarely pure substances. Atoms are building blocks of our world. Electric current results from moving charged particles. Salts are raw materials for many useful substances. Insulators do not contain charged particles which can move. Voltage can cause movement of charges. Voltage and resistence determine electric current. Electricity and energy. Heat causes alteration of the state of substance. Heat as a form of energy. Changes of heat energy. Cold is lack of heat. The sun determines our climate and is our main source of energy.

Year 4: *Living with chemistry.* Water as a chemical. Acids and bases in everyday life. The dosage is important — chemicals in everyday life. Sodium chloride: in nature and as the product of a chemical reaction. We all breathe air. Raw materials in working life. Synthetics and macro molecules. Ethanol and acetic acid. Nutrients in our food. Chemistry and cleanliness. Natural and chemical products complement each other. Economic use of energy and natural resources.

The Advanced Chemistry course in the AHS for students aged 16–18 is:

Chemistry and the substances of our environment. Atoms, the Periodic System and the signficance of models. Properties of substances in relation to chemical bonding. Why chemical reactions occur and how they may be controlled. Chemical reactions and energy — the importance of entropy. Fossil energy resources (introduction to organic chemistry). Nature and industry can change organic compounds (mechanisms of organic reactions). The significance of modern analytical measurement techniques for health and environment. Renewable resources — food for mankind (introduction to biochemistry).

21

Chemical Industry and a High School Chemistry Curriculum in Israel

R. BEN ZVI
Weizmann Institute of Science, Israel

The educational system in Israel has gone through a series of changes during the last few decades, changes which were at least partly dictated by demographic and socioeconomic factors. As both the changes in the system and the factors which caused them, have a direct bearing on science teaching, it seems appropriate to describe them briefly as a background to current trends in the teaching of chemistry.

In the past, the academic high school could be regarded mainly as a bridge to higher education because a comparatively high percentage of the students who went through it found themselves eventually in the university. Students in high school could choose between science oriented or humanistic oriented streams and many of those who chose the physical science or biological science streams did so in order to prepare themselves for future careers in the sciences.

The rapid changes in the population of Israel due to massive immigration from all over the world caused a change both in the structure and in the orientation of the school system. The students, nowadays, vary very much in their abilities, interests and needs, and therefore the high school is looked upon more as a tool for the preparation of future citizens than as a stepladder to university. In accordance with this, the rigid streaming system has been changed into a more flexible one of creditation enabling each student to choose subjects according to his or her inclination.

The high school, nowadays, is a 6 year school composed of 3 years junior and 3 years senior high school. In the junior high school (age 13–15) all students study a combined physical science course. In the first year of senior high school, all students have to study each of the science subjects (chemistry, physics and biology) and afterwards, for the last 2 years, students are free to choose whether, and to what extent they want to continue their science studies.

Chemistry programmes in Israel

The first new programme implemented in Israel in the mid sixties was a translation of the CHEMStudy programme developed in the US. That programme which is concept-oriented, based on the structure of the discipline was suitable for students who have true interest in the sciences. It is, however, not suitable for the present high school population.

The currently used textbooks "Chemistry for High School" were written in 1972 and were originally intended for students in the science streams. In order to make it useable in the new school system extensions have been added for the various student populations.

The first set of extensions (eleventh grade) is taught to students who do not intend to specialise in chemistry but wish to continue their studies in this area beyond the tenth grade. The second set, on the other hand, is planned as additional material for students who have chosen chemistry as one of their main topics of study. Although the two sets differ both in the scope of subjects presented and in the depth in which they are taught, they are similar in the approach which is interdisciplinary and in the attempt to widen the scope of chemistry.

Kempa[1] suggested that there are at least six facets of chemistry each of which should be accorded a legitimate place in the chemical curriculum for the secondary level. The first two he mentions are those related to the structure of the discipline while the four additional ones are:

a. the technological manifestation of chemistry;
b. chemistry as a "personally relevant" subject;
c. the cultural aspects of chemistry;
d. the societal role and implications of chemistry.

In each of the extensions as many as these aspects as possible have been introduced as will be exemplified in the following section.

Chemical industry and high school chemistry

The main objective of the two units on chemical industry (see Fig. 1) is to provide a link between chemistry as taught in the classroom and the outer world. More specifically the intention was[2]:

a. to demonstrate how basic chemical principles are applied on an industrial scale;
b. to demonstrate the importance of chemical industry to society and to the economy;
c. to develop a knowledge of the technological, economic and environmental factors involved in chemical industry;
d. to investigate some specific problems faced by the local chemical industry.

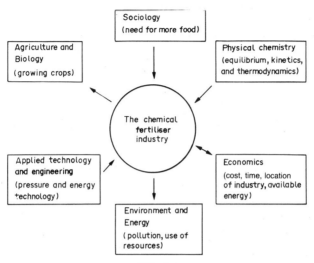

FIG. 1

The first unit — "The Industry of Fertilisers" is based on very basic principals and ideas which were taught in the tenth grade. Aspects of equilibrium and kinetics are, therefore, introduced and applied to consideration of the industrial production of ammonia, nitric acid and fertiliser. These theoretical ideas are integrated with various aspects as presented in Fig. 2.

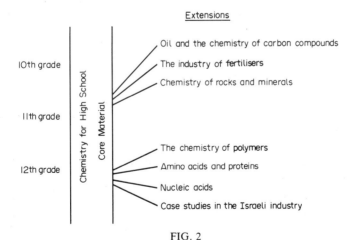

FIG. 2

The second industrial unit "Case Studies in the Israeli Industry" includes the following issues[3]:

   a. relevant examples of chemical applications in industry;
   b. social and economic issues;
   c. a balanced picture of ecological problems;
   d. examples of decision making;
   e. an interdisciplinary approach to problem solving.

The three case studies chosen for the unit are:

a. *Copper Production in Timna* which introduces problems involved in the construction of a chemical plant, copper production, cost factors and the considerations that led to shutting down of the mine.

b. *Plastics: the manufacture of polyvinyl chloride.* This case study deals with the chemistry of the monomers and the polymerisation process and also with the ecological problems and health treatment to workers in this industry.

c. *Life from the Dead Sea.* A study of the production of bromine and its compounds and their uses. The problem of extracting bromine from its salt solution (Dead Sea water) is discussed. Students perform simple experiments and using given data, determine the availability and cost of raw materials, the energy needed, the safe disposal of byproducts and technical problems involved in scaling up to full production[2].

The implementation of interdisciplinary units of the nature described above, causes many problems. The chemistry teachers who were educated in the University consider themselves as experts in their own field, and feel very insecure and therefore reluctant to be involved in discussions of social and sometimes even moral issues. Very intensive inservice teacher training courses were, therefore, designed in order to introduce the specific aspects of these units[3].

In these courses teachers were introduced to the structure of the Israeli chemical industry and to specific elements of the industrial process. They discussed the difference between small-scale laboratory experiments and large-scale industrial production, learned how calculations of material and energy balances are made and were introduced to elementary skills such as how to read flow charts or how to run small-scale industrial type experiments. More general issues were also brought up such as cost variables and the balance between supply and demand.

The two units described above, and the other extensions mentioned, have been taught successfully for the last couple of years and helped in showing the relevance of chemistry to everyday life. But these extensions, which were developed with this specific objective in mind, are not the only points of contact between classroom chemistry and the outer world. Throughout all the years of study, starting from the

8th grade course, national, social, economic and industrial issues are brought up as will be illustrated by the next example.

The Dead Sea, a source for chemistry teaching

Israel is very poor in raw materials and the only natural source is Dead Sea with its high concentration of various minerals. Figure 3 shows how this source is used throughout the grades as a motivation factor for various topics in chemistry. One example will serve to show how this is done.

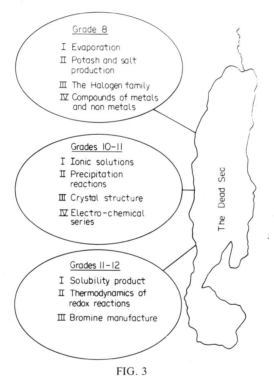

FIG. 3

One of the topics taught in the 10th grade concerns ionic solutions and precipitation reactions. The idea that sometimes when two ionic solutions are mixed, a precipitate is formed, is introduced via a nation-wide project which is now in the planning stage — the Mediterranean — Dead Sea Project. In this project it is planned to dig a channel connecting the Mediterranean Sea (sea level) with the Dead Sea (-400 metres) and use the height difference for the production of energy. One of the many problems, discussed much in the daily newspapers is the possibility that in the specific conditions prevailing, $Ca^{2+}_{(aq)}$ ions from the Dead Sea (16.9 g/litre) will form a precipitate with $SO_4^{2-}_{(aq)}$ ions from the Mediterranean water (3.6 g/litre). A simulation experiment in which samples from the two sources are mixed, starts the chapter dealing with precipitation reactions.

References

1. Kempa, R. F., Developing new perspectives in chemical education, a lecture presented at the 7th International Conference of Chemical Education, Montpellier, France, 1983.
2. Hofstein, A. and Nae, H., Chemical case studies: science — society "bonding". *The Science Teacher*, **48**, 52–53, 1981.
3. Nae, H. and Hofstein, A. "Inservice teacher training for the incorporation of new interdisciplinary courses on the chemical industry"; In Tamir, P., Hofstein, A. and Ben Peretz, M. (eds.) Preservice and Inservice Education of Science Teachers, Rehovot, Balaban, 1983.

22

Introducing Industry into Science Teaching

Riccarton High School, Christchurch, New Zealand

"Science and technology have never faced as serious a challenge as the one posed by the pressing problems of mankind today."[1] At the fourth International Conference on Chemical Education in Ljubljana in 1977, C. N. R. Rao made it clear that chemistry teachers have a crucial part to play in the search for solutions.

At the same conference G. C. Pimentel proposed four goals for both secondary and introductory college level chemistry programmes.[2]

1. To encourage rational thinking and to give practice in problem solving.
2. To promote scientific literacy in future citizens.
3. To provide future citizens with a readiness for technological change.
4. To provide a foundation for a professional career in science or some other technical subject based on chemistry.

The first three of these goals are appropriate for all students, whatever their abilities and aspirations and it is these three, together with their social and economic implications, which challenge the adequacy of these chemical curricula which look no further than the laboratory and the textbook of facts and theories. With these goals in mind, a narrowly academic approach is no longer adequate. Wider goals for school chemistry acknowledge that the world we live in is increasingly dependent upon the chemical industry and on industrial research and its applications.

Students naturally expect to receive logically constructed science courses presented in a competent way, but today's students require more than this. They should be offered courses which relate in two ways to the world outside the classroom; the course should include both demonstrations of the benefits chemistry brings to society and clear illustrations of the problems that can accompany or follow developments in chemical industry.

Many teachers feel more comfortable teaching facts and concepts. They also appreciate the limitations in their own abilities in assessing the impact of technology on people. Science is neutral with respect to value judgements and attitudes toward its social implications. It is important that the distinction be made between what chemistry can do and what it cannot do; between its capacity to provide facts, data, theories and speculations and its inability to provide solutions to those difficult problems involving conflict of interest in the development of human and economic resources.

If teachers do not recognise the importance of the chemical industry, they imply that it is unimportant. Even more serious consequences follow if they do not attempt to combat the excessively negative attitudes towards chemistry sometimes promoted by the media. When such an attitude arises from a misunderstanding or a misrepresentation of the chemical facts, a clear explanation becomes a powerful illustration of the need for informed appraisal of such news reports. We need to emphasise, however, that there is a price to be paid for development. Bigger and better yields from crops, alternative energy sources, more effective health care, faster and more comprehensive communication and other forms of technological impact do bring with them negative effects. These may be obvious, such as the pollution of the atmosphere or waterways or they may be less apparent, as when scientific knowledge is used as a means of power and control. The thoughtful use of selected examples from chemical technology is one of the best methods for directly promoting the first three important goals. New Zealand is one of a number of places in which attempts are being made to promote these aims. The following guidelines may be helpful.

I.   Programme aims should include industrial processes

In lists of general aims, specific reference to the place of chemistry in society should be included. Such a statement is:

> "To lead students to understand the role of chemical science in the society in which they live and its importance in placing in proper perspective the current conflict between technology and conservational restraint. Further, to introduce students to some of the economic considerations which influence the development of industries and the use of alternative materials and processes".[3]

Particular courses of study should refer to chemical processes that are related to national and local resources. For example, the science programme in New Zealand intended for students aged about 15 years includes the following:

> "Extraction of iron and aluminium from their principal ores as carried out in New Zealand; the chief uses of these metals in industry.

Chemical aspects of the manufacture and use of sulphuric acid and superphosphate.

Petroleum, natural gas, liquid propane gas and compressed natural gas as important mixtures of alkanes.

Manufacture of ethanol, from ethane and by fermentation, its combustion and its use as a fuel. The manufacture of methanol from natural gas and its uses."[4]

These are economically important industries in a relatively isolated and predominantly agricultural country with very limited indigenous mineral resources.

Examination prescriptions at senior levels in New Zealand schools emphasise chemical principles. Applications of the principles are only specifically referred to when they are large-scale processes of major importance. For example, the topic *equilibrium* is expanded to include the effects of change of temperature, pressure and concentration of components and the composition of the reaction mixture. The syllabus then refers to "the application of these effects in industrial chemistry illustrated by the Contact Process for sulphuric acid manufacture and Haber Process for ammonia synthesis".[5]

National programmes cannot be expected to take into account specific local cottage industries or particular community concerns. Each teacher should be prepared to take the initiative in searching for examples of chemical processes going on in his or her immediate environment.

II.   Select what is relevant

Most people do not view chemistry in the same way as those who work in research laboratories or who teach it. They do not see it as a myriad of chemical facts or as a series of big abstractions. If chemistry is to be part of general education, with major economic and social importance, then criteria for selecting material for courses are needed.

(i)   *Emphasis should be given to what chemists do and not merely to what chemists know.*
Making compounds, identifying and modifying substances to give them more desirable properties, finding alternative materials for substances in short supply and inhibiting or accelerating chemical processes are all things chemists do. Industrial processes demonstrate these.

Approaches along these lines have the added advantage of keeping the laboratory, rather than the classroom, central in the teaching of chemistry.

(ii) *Make sure that small-scale local industry is not overlooked.*
Soap making, choosing fertilisers, stemming corrosion, treating water, fermentation and other activities like these provide examples close at hand which relate to chemistry which students will meet in the laboratory.

(iii) *Make clear that there are important implications when the scale of a chemical reaction is altered.*
The large-scale chemistry of industry requires different methods from those of the laboratory bench.

(a) *Choice of starting materials.* Cheapness and availability are likely to be more important to the industrial chemist than chemical convenience. Methanol manufacture in New Zealand uses a catalytic reaction of methane and steam. A laboratory chemist would be most unlikely to consider making methanol this way.

(b) *Industrial processes are often continuous flow systems.* They are simpler to engineer and control. The bench chemist usually requires only a single preparation.

(c) *The energetics of industrial processes are frequently extremely important.* Laboratory chemists find it convenient to use the bunsen burner, hot plate or cooling bath to control the temperature of the reaction vessel. In large-scale production heating the reactants or getting rid of the heat produced in an exothermic reaction is a major problem for chemical engineers. The energetics of a chemical reaction may influence critically the yield of desired product. For example, in the Contact Process for the manufacture of sulphuric acid, sulphur is burned to sulphur dioxide, oxidised catalytically to sulphur trioxide then combined with water. Each of these three stages is exothermic, but even with a catalyst the conversion to sulphur trioxide only goes at a significant rate above $420°$. However, conversion from SO_2 into SO_3 falls off rapidly above $500°$. Optimum production of sulphur trioxide is only achieved by carefully balancing the opposing kinetic and thermodynamic effects. This requires good plant design to give precise temperature control if the maximum yield of product is to be obtained.

III. Acknowledge environmental issues

However important economic growth and the consequential material benefits are particularly to developing countries, there is a growing concern in today's world that:

(i) *Global resources are diminishing.*
Not only are the stocks of many metal minerals being rapidly depleted but the use of non-renewable energy resources in the recovery process may have long term consequences which will be just as serious. New Zealand's steel industry depends upon the recovery of iron from beach ironsands. This is available in sufficient quantity to last for a thousand years at present production levels. However, coal is also required for the reduction process at the rate of a million tonnes a year. It is obtained from the open-cast mining of what has been good farmland. Big holes are going to be a permanent reminder of the price paid long before the ironsand runs out.

(ii) *There is a need to conserve resources.*
The value of recycling is readily appreciated. It is worth also pointing out that not only the materials but also energy may be saved by recovery programmes. It takes about 1.6×10^{10} kJ of energy to produce a tonne of aluminium. Recycling the same amount of metal requires less than one thousandth of that quantity of energy.

(iii) *Pollution needs to be monitored and controlled.*
The production of aluminium by the Hall-Heroult process requires the electrolysis of alumina dissolved in molten cryolite, Na_3AlF_6. At the operating temperature of nearly 1000° some of the fluoride is given off from the smelter pots along with carbon monoxide and carbon dioxide. The plants operating in Southland, New Zealand, are designed with multi-cyclone filters to spin out fluoride dust particles. Waste gas is then ejected from a 137 metre stack to disperse into the atmosphere. Farms in the area are regularly monitored and the health of plants and animals checked. So far fluoride levels have been well within standards laid down by Government authorities.

Evaluating the effectiveness of antipollution measures is likely to be beyond the competence of the teacher without a background in economics, local social issues and national aspirations. However it is still a legitimate function of the science teacher to point out the environmental impact which chemical industry frequently makes and also the steps taken to minimise it.

IV.   Additional resources are needed

(i) *Help from industry.* Teachers need to known only general principles and details of the main features of industrial processes. Technical data particularly of the kind likely to provide a competitive advantage is not required. Even small companies should have designated staff able and willing to provide basic information and pictures. Some industries lend films, videotapes and display material including samples of the raw materials and the products.

(ii) *Joint ventures* with practising chemists are one of the best methods, not only for developing resources but also for enhancing the professional development of teachers. One such project in New Zealand involved the preparation of a substantial book[6] covering as many different chemical industries as was possible. The main features of each industry were summarised by a teacher assisted by a chemist or chemical engineer engaged in the process. Publication was arranged by the national professional organisation of chemists. This resource book provides not only teaching information but also background material to deepen the teacher's understanding.

(iii) *Simulation situations and case studies* are being used in several countries and are reported to bring about significant changes in the attitudes of students toward chemistry. Using techniques such as role play, structured discussion, group decision making and competitive planning, student involvement is reported to be high.[7] Examples of some games and simulations are given on page 159 onwards.

The basic aims of structured material have been precisely stated.[8]

(1) The studies should be based on clearly defined chemical processes which are related to the "core" chemistry course.

(2) The basic technology of the industrial process described must not be too difficult to understand.

(3) The economic and social aspects must be clear cut and easy to explain.

(4) It should be possible to include a discussion of topics such as raw materials, manpower, energy requirements, waste disposal etc.

Case studies based on criteria like these have a high potential in motivating and improving students' interest in chemistry.

V.   The method of assessment affect what is learned

Students learn largely what they expect to be examined on. In introducing a component of industrial chemistry into high school programmes there is a risk of overloading syllabuses. Technical data readily lends itself to rote learning. Multiple choice type testing, unless skilfully used, can be a superficial method of assessment.

Care must be taken that the focus in assessment remains on the principles of the processes rather than a multitude of technical details.

It would also be counter-productive if time which might otherwise be spent on practical work becomes diverted into an excessive commitment to the exploration of the chemical industry.

VI. Conclusions

Introducing industrial processes into high school chemistry programmes promotes the wider subject aims, and opens a window to the world outside the school.

Examples of selected processes should be written into syllabuses and teachers should be provided with resources from industry from which they may select information to illustrate chemical concepts.

Teaching programmes should not be overloaded with factual information.

It is a teacher's responsibility to draw attention to environmental issues associated with chemical processing. However, they should accept the constraints of their own backgrounds in attempting to evaluate conflict of interest particularly where social issues and values are involved.

References

1. Rao, C. N. R., *Chemical Education in the Coming Decades*, International Symposium on Chemical Education, Ljubljana, ed. A. Kornhauser, 1979; 9–16.
2. Pimental, G. C., *Ibid*; 108.
3. Watts, D. W., *Objectives for the Royal Australian Chemical Institute, Chemical Education — Chemistry Across the Secondary-Tertiary Interface*, Royal Australian Chemical Institute, 1979; 20.
4. *School Certificate Science Prescription*, Department of Education, Wellington, New Zealand.
5. Chemistry for University Entrance, *University Grants Committee Handbook* 1984; 193.
6. Packer, J. E., *Chemical Processes in New Zealand*, New Zealand Institute of Chemistry, 1978.
7. Laslett, R. L. *Resources for the Teaching of Industrial Processes, Chemical Education — Chemistry Across the Secondary-Tertiary Interface*, Royal Australian Chemical Institute, 1979; 28. (A comprehensive list of resources in English).
8. Nae, H. Hofstein, A. and Samuel, D., Chemical Industry, *J. Chem. Ed.* 57, 1980; 366–8.

23

Introducing the Chemical Industry into the Science Curriculum in Papua New Guinea

A. C. W. PONNAMPERUMA
Bena Bena High School, Papua New Guinea

W. P. PALMER
Goroka Teachers' College, Papua New Guinea

Papua New Guinea is a young, developing nation, achieving independence in 1975. Industries which can be classified as chemical industries by western standards, are almost non-existent in Papua New Guinea. As yet, most of the raw materials it produces are exported to other countries to be processed or manufactured into useful products. However, some factories have already been built; more will be built in future. At this stage of the country's development, we can consider the mining industry and the processing of local primary products which utilise the basic techniques of chemical technology. Amongst those industries which are in successful operation are copper and gold mining, beer brewing, the production and refining of sugar, ethanol production from sugar wastes, food processing, nail and wire products, polyurethene foam products, match manufacture, and many others. In recent times the export of copper concentrate from the Bougainville Copper Ltd. project alone, contributes about a third of the country's foreign exchange earnings.

In this paper, we explain how the lower secondary school pupils are introduced to Chemical industry in Papua New Guinea, with particular reference to the mining and extraction of copper on Bougainville island.

Science education in the secondary school system

The development of secondary education in Papua New Guinea dates from the late 1950s, and science became an integral part of the secondary school curriculum in the early 1960s when secondary education was mainly confined to the lower grades. At present, the secondary education system in Papua New Guinea, consists of two stages:

(i) The lower secondary level provincial high schools for young people aged 13–16 (grades 7–10). There are about 110 such high schools.

(ii) The upper secondary level national high schools for students aged 17–18 (grades 11–12). There are four national high schools.

Science is studied by all students in grades 7 to 11. In addition, there are two applied science subjects available in provincial high schools, namely, agriculture and home economics.

In national high schools, at grade 12 about 60% of the students take up a major science course which can lead to further studies in science. Another 20% study minor science courses and the remainder do not study science. Major courses in science are centred around chemistry, physics and biology courses which are taught separately, while the minor courses in science consist of modules selected from three themes — human biology, evolution and technology.

Science curriculum in provincial high schools

The early syllabuses used in science teaching were developed overseas, in Australia and Sarawak, but work began on a secondary science syllabus designed for Papua New Guinea in 1965. The original science curricula developed in Papua New Guinea laid much emphasis on "academic science" similar to the formal science curricula used in developed countries.

The emphasis in science curriculum development in Papua New Guinea has been towards a "general" or an "integrated science" education for all, at the provincial high school level. The secondary students are introduced to more specialised branches of science such as chemistry, physics and biology in grade 12 at the national high school level. The provincial high school science course consists of 22 self-contained units:

Grade 7:

　　Unit 7.1 – Introduction to Science
　　Unit 7.2 – The Sun and the earth
　　Unit 7.3 – Matter
　　Unit 7.4 – Living and non-living things
　　Unit 7.5 – Heat energy
　　Unit 7.6 – Electricity

Grade 8:

　　Unit 8.1 – Changes in matter
　　Unit 8.2 – Energy in living things
　　Unit 8.3 – Electricity 1
　　Unit 8.4 – Force, work and energy
　　Unit 8.5 – Growth and reproduction
　　Unit 8.6 – Science in Society

Grade 9:

    Unit 9.1 – Air around us
    Unit 9.2 – Electricity 2
    Unit 9.3 – Communication
    Unit 9.4 – Traditional technology (optional)
    Unit 9.5 – Ecology
    Unit 9.6 – Our body

Grade 10:

    Unit 10.1 – Chemical technology
    Unit 10.2 – Light
    Unit 10.3 – Microbiology
    Unit 10.4 – Geology

There are only three units on chemistry. These are the units on matter and particle theory (grade 7), changes in matter (grade 8) and chemical technology (grade 10).

There are many reasons given for keeping the chemistry content low at provincial high school level. Firstly, the schools are not well equipped and the majority of science teachers at provincial high schools are Goroka Teachers' College graduates who are not at present trained to teach chemistry, other than that contained in the provincial high school syllabus. Further, chemistry is an "abstract science" which demands more "formal thinking". Piagetian research done with national high school students and the preliminary year students from the University of Papua New Guinea has shown that even at a more mature age than the 13 to 16+ age group at provincial high schools only a small minority of Papua New Guinea students are capable of such thought.[1] Since a large majority of students at provincial high schools are still at the "concrete stage" of thinking the introduction of highly abstract and conceptually difficult materials has been considered irrelevant for Papua New Guinea.

In the chemical technology unit taught at grade 10, students are introduced to industrial aspects of chemistry related to Papua New Guinea. This unit deals with industrial aspects of copper mining and extraction, the chemistry involved in the traditional methods of converting limestone to lime, the use of traditional and modern fuels in Papua New Guinea, and the chemistry involved in water treatment.

The Bougainville copper mining project has brought about significant changes in the country's society and economy, and as such, the section on copper attempts to create an awareness and a basic understanding of the uses of copper and the chemical technology involved in its mining and extraction. In this section the students are introduced to the uses of copper based on its properties as an excellent conductor of heat and electricity and also on its resistance to corrosion. Students also investigate the physical and chemical properties of copper ore and copper concentrate samples.

The students are then introduced to the preliminary steps involved in the mining and crushing of copper ore. Simple models of ball mills are used to demonstrate the action of the ball mills used in crushing the ore to fine particles. The separation of copper ore from waste by oil flotation is also discussed. The students also investigate the presence of copper in Bougainville Copper concentrate using a test tube with ammonium hydroxide solution and the flame test for copper ions. The reduction of copper ore and the electrolytic refining of copper are also investigated.

Although introducing the students to chemical industry in their own country has its advantages, the students do encounter problems in understanding the chemical principles utilised in these industrues. There are a number of reasons for these problems. Firstly, the lack of sufficient prior knowledge of the theoretical background of chemical reactions (for example, oxidation, reduction, electrolysis, etc) poses a major problem of understanding for students. Secondly, the lack of relevant supplementary reading materials suitable for Papua New Guinea students for whom English may be a third or fourth language also aggravates learning difficulties for students. Thirdly the provincial high school students, who are mainly "concrete thinkers", have problems visualising the real "industrial" situation, although models are used in some cases (for example, a ball mill). Lastly the geographical distribution of the provincial high schools in the country, coupled with the high costs of travel make it almost impossible for students to visit an industry such as the Bougainville copper project, to see industry for themselves.

Recently, one of the authors (WPP) designed a series of slides of the Bougainville copper mines showing the various stages of mining, crushing, and preparation of copper concentrate for export, which could be made available to teachers as a visual aid to help overcome some learning difficulties, arising from the students' inability to visualise the stages of the mining industry.

The provincial high school science curriculum advisory committee recognised some of the problems of teaching and learning this unit. They recommended the following aims and objectives:

Aim: 1.   To improve general knowledge and understanding of chemistry.

2.   To study the properties, chemical reactions and uses of familiar naturally occurring substances.

TOPIC	SPECIFIC OBJECTIVES
Extraction and Production	– to know that less reactive metals are more easily extracted from their ores than reactive metals; electrolysis and reduction.
Properties and Uses	– to know the physical properties used to distinguish between metals and non-metals; Cu, Fe, Al, Pb, metals and their alloys. – to know the uses of copper alloys; fuels. – to understand the national and international economic importance of copper and crude oil.
Chemical Reactions	– to understand that metals have different chemical activity and this determines rate of corrosion, natural occurrence and ease of extraction.
Extraction and Production	– to understand the processes involved in the production of lime; the extraction of copper; the production of fuels from wood and crude oil – to know that plastics and other useful products can be obtained from crude oil.
Chemical Reactions	– to know that in a chemical change substances form new substances; changes in combustion of hydrocarbons; neutralisation of acids by bases; reaction between acids and carbonates; smelting of copper; decomposition of calcium carbonate. – be able to use simple word equations.
Natural Substances	– to know that rocks contain useful substances; to understand the role waters play in dissolving naturally occurring substances; to know reactivity and solubility determine the form they occur.

The revised Chemical Technology Unit is to include more emphasis on chemical concepts which might help overcome some of the deficiencies of the existing unit. However, as Wilson and Wilson have pointed out, the success of any revision will depend to a large extent on the use of teaching strategies which will deliberately encourage the development of formal thought in students, and the level of understanding of the chemical concepts by the teachers themselves.[1]

Reference

1. Wilson, A. and Wilson, M. (1983) "The extent and development of formal operation thought among national high school and preliminary year students in Papua New Guinea". Report to Department of Education, Waigani, Papua New Guinea, pp. 2–25.

Background Reading

Haddon-Smith K. (Aust) Pty Ltd, "Bougainville Copper", Bougainville Copper Ltd, Printed in Singapore by Toppan Printing Co. Pty Ltd., pp. 4–17.
C.R.C. in Copper "Minerals to manufacturing", Printer C. R. A. Limited, 55 Collins Street, Melbourne 3001.

24

Introducing Chemical Industries Into the Secondary Curriculum in the Philippines

M. C. TAN
Institute for Science and Mathematics Education Development, Philippines

In the Philippines, the secondary chemistry curriculum has been revised to provide opportunities for students to realise that principles studied in class are in fact part of a real-life situation. The environment and the local chemical industries provide an interesting and useful vehicle for demonstrating these applications and for introducing relevant issues for students in secondary school.

The Philippine government started in 1975 a massive text-book development programme written by Filipino authors for Filipino students. The University of the Philippines Institute for Science and Mathematics Education Development was officially designated to undertake the development of science and mathematics textbooks for elementary and secondary schools. *Chemistry In Our Environment* was produced under this project and it has been the core textbook in use by Grade 9 students in our country since 1981. It presents basic chemistry principles and their applications suitable for terminal and college-bound high school students.

Because of the varying background and intellectual development of our students a flexible curriculum was developed that answers the separate needs of the college-bound students and those who drop out of high school in the different communities. The curriculum suggested the core topics found in the textbook *Chemistry In Our Environment* and enrichment lessons of different topics which come in the form of self-contained learning units or modules. The flexible curriculum demonstrates that chemistry is much a part of our lives, all of the time, and not simply what is done in the school laboratory. It uses experiments to pose problems, to develop patterns and to lead to general principles. This approach familiarises students with common chemicals and their reactions, and takes them from concrete experience to abstract ideas. The influence of chemical industry and technology in their daily lives emerges naturally from the experiments and discussions used to illustrate the various topics. The features of the curriculum materials are described in the following paragraphs.

The text

Content is relevant to students' experiences and the local environment

The text emphasises chemical concepts and principles applied in local industries, agriculture, and daily living. For example, the chemical principles involved in the purification of water for household use as practised at the Balara Water Purification Plant in Quezon City are taken up in the lesson on liquid mixtures. The discussion on land environment extends to minerals found in the Philippines and includes the latest mineral map of the country. The importance of pH in the growth of local plants in industries and in human body processes is brought out in the lesson on electrolytes. Temperature and its effects on reaction rate leads to the chemistry of fire control.

In order to make the students understand better the scientific phenomena affecting them, topics are included on environmental pollution, conservation, nutrition and health. For example, the algal bloom episode in 1975 in Laguna de Bay where millions of milk fish (*bangus*) died, is discussed in relation to man's agricultural and industrial activities. The current problem of smoking among students is approached by showing in an activity the poisonous chemicals inhaled from cigarettes. A background of drug abuse is also included.

In some cases, environmental phenomena introduce the concepts as in the problem of corrosion, which brings in the concept of oxidation-reduction reactions.

The use of local materials in the activities is recommended

The new textbook recommends minimum purchase and use of expensive chemicals. Many activities in the textbook use local materials found in the kitchen or the local stores. Activities on the recycling of waste materials into useful chemicals are also included. Some chemicals prepared in one activity can be used in subsequent activities. For example, in the activity on the separation of substances in solid mixtures, MnO_2 recovered from spent dry cells is used. Most activities require only the use of bottles and very simple set-ups. The only piece of "sophisticated" equipment required is an electrical conductivity apparatus to be prepared by the students in one activity and used later in five other activities.

The language used is simple

Geographical and historical factors contribute to the language difficulty of the students. Most students have difficulty comprehending not only technical words, but also non-technical words used in science. In 1975, the

WISP study (Words in Science Philippines) listed words according to their difficulty range and this was utilised by the chemistry team during the writing phase of *Chemistry In Our Environment.* Writers used simple words whenever possible. When the use of a difficult word could not be avoided, the meaning was given in parentheses. Ample illustrations and pictures are provided in the materials to clarify and simplify instructions and discussions.

The content is adapted to student's cognitive level

Chemistry as a discipline requires a lot of formal thought on the part of the students. Most of its basic concepts are abstract. To suit the learning capacity of the target users at their level on cognitive growth, the lessons are presented and sequenced starting with the macro world gradually leading down to the molecular level and finishing off at the micro level. Concrete concepts are presented before the abstract ones. This is one of the striking contrasts between the curriculum materials the UPISMED team has developed and the other materials available. Other textbooks traditionally start with atomic structure and students are then faced with subject matter beyond their comprehension. This scares them away from chemistry.

Learning by doing is emphasised

Activities are integrated with the text. The observations and conclusions drawn from the activities are necessary links in the development of the concepts. The book also shows the application of theory to practical situations and provides students with concrete experiences of abstract principles. It ensures active participation by students in the learning process. Besides the activities, questions are interlaced in the text to develop in the students inquisitiveness and critical thinking. *A Handbook of Laboratory Procedures and Techniques*[1] was written as a reference book for teacher and students.

Integration with other sciences is shown

Whenever appropriate chemistry is related with other fields of science. For example, the biological as well as climatic effects of pollutants are discussed in the lesson on mixtures. The physics of heat, the kinetic molecular theory and atomic structure are also integrated. Concepts in earth science such as delta formation and weather disturbances are likewise discussed in relation to chemical phenomena.

The teacher's edition is included

The teacher's edition of the text can help the teachers who have inadequate training in teaching chemistry. All of the students' text is contained in the teachers' edition version together with some instructions for teachers.

The enrichment materials

To supplement the core textbook and update references used in schools, topics important for college-bound students like *Chemical Language* and *Chemical Arithmetic* which are treated superficially in the text are discussed with more depth in the enrichment materials. For terminal students, applied modules that provide knowledge and skills for simple home and community activities are available. Many modules illustrate the importance of local chemical industries to society and to the economy of the country and include discussions of the relevant economic and environmental constraints (Table 1).

Mostly developed under the ongoing Science and Technology Education in Philippine Society Project (STEPS), the applied enrichment materials help develop some skills in utilising indigenous resources, as well as skills in inquiring and decision making. These modules also help the students recognise the finite nature of the Philippines' and the world's natural resources, as well as how to conserve and use wisely local resources for the sake of the country's present and future needs. The different enrichment materials explain basic chemistry concepts and use some simple science processes and mathematical skills involved in what is grown in and taken from the local environment and in what is made in the country. Note that most of these modules are agro-based. There are others being developed.

The modules on plastics, wood chemicals, sugar, detergent, food processing, nuclear and cement present the actual processes followed in industry. The science behind the processes are emphasised. However, the activities done by the students use locally available materials and in micro amounts. The other modules answer the needs of specific types of communities.

How the curriculum materials are used

Each chapter in the text is divided into two or three lessons, a lesson constituting a learning unit of several related concepts. This arrangement makes the book flexible in terms of content and teaching techniques. For example, if the teacher is pressed for time, he or she may omit a lesson, provided it is not a prerequisite to succeeding lessons. Also, if desired, a shift from class pacing to individual or small group pacing can be made for a series of lessons. In the teacher's edition, a flowchart gives the sequence of lessons along with some supplementary material.

Table 1. List of Chemical and Chemical Related Industries and the Available Enrichment Modules Under Each Category

Industries*	Enrichment Modules Available
A. Water Conditioning and Environmental Protection	A1. Environmental Series: Air, Water, Soil 2. Potable Water From Sea Water by Solar Distillation 3. Geothermal Energy: It's Chemical and Thermal Effects 4. Making a Water-Sealed Toilet
B. Energy, Fuels, Air Conditioning and Refrigeration	B1. The Refrigerator: How it Works 2. Geothermal Energy: An Introduction 3. Electricity From Under the Ground 4. Charcoal From Coconut Shells 5. Waste Not! Want Not!
C. Portland Cement, Calcium and Magnesium Compounds	C1. From Rocks to Cement 2. Hollow Blocks From Soil and Rice Hulls
D. Chlor-Alakli Industries	D1. Making Lye From Wood Ash
E. Nuclear Industries	E1. Go Nuclear?
F. Food Production/Food Processing	F1. Food Preservation Series 2. Corn In Your Backyard 3. Nata de Arroz 4. Collecting and Handling Bangus Fry 5. Fish Modules Series 6. Money in Rabbits 7. Preparation of Fish and Shrimp Bagoong
G. Agrichemical	G1. Pesticides 2. Medicinal Vegetables and Spices 3. Twenty Common Medicinal Plants And How To Use Them
H. Soaps and Detergents	H1. Detergents: For Whiter Clothes or Greener Lakes 2. Making Soap From Used Oil
I. Sugar	I1. From Cane to Sugar 2. Waste Water Treatment of the Sugar Industry
J. Wood Chemicals	J1. Cocowood Treatment by Double Diffusion Method
K. Plastics	K1. Plastics In Our Lives

*Classification of industries adopted from Shreve and Brink. *Chemical Processes Industries.* McGraw-Hill Book Co., 1977.

Although the core text is used by all government schools, the enrichment materials are not. The teachers only choose what is appropriate for their students and community. For example, fishing communities can use the series of modules on Fish. Originally, these modules were prepared to answer some needs of San Salvador, a fishing community in Iloilo, Panay Island, the site of the UNICEF-supported project "Survival of the Family" which aims to improve the living conditions in a community through learning situations in school[2], but they can be adapted by other fishing communities. The set of three fish modules describes common fish in the Philippines, ways to conserve marine resources, causes and effects of water pollution and its effects on fish life.

It is hoped that this curriculum will be carried on in the tertiary level as a formal linkage between industry and education as recommended in a survey undertaken by the Philippine Council for Industry and Energy Research Development. The study considered the gap between manpower needs of industry and the output of the schools. As industry in the Philippines approaches the level and importance of agriculture as a producer of goods, a user of human resources, and a contributor to the economic development process, this problem becomes crucial to its ability to realise plans and successfully apply strategiers for national progress and competitiveness in international markets. The study locates industry in the Philippine economic picture and examines its needs and capabilities particularly in manpower resources; it provides information on the relative importance of the different manufacturing sectors in terms of economic activity; it projects the human resource needs of industry up to year 2000; it examines the engineering and technical curriculum of the Philippine educational system; it states the expectations of industry from graduates of engineering and technical courses, and identifies possible areas of mutual co-operation between industry and the schools to strengthen the practicum stage of the learning process; it identifies specific requirements of industry that should be considered in curriculum construction; and it prepares an action programme to implement the new educational scheme.

References

1. P. da Silva and M. C. Tan. *Handbook of Laboratory Procedures and Techniques*. Q C. Alemars Phoenix Press. 1979.
2. ISMED Monograph No. 19. Case Study 1: The Concept of Health and Environment Among Children of a Fishing Community. 1981.

25

Industry and Technology: the CHEMCOM Philosophy and Approach

S. WARE
American Chemical Society, Washington, USA

H. HEIKKINEN
University of Maryland, USA

W. T. LIPPINCOTT
University of Wisconsin, Madison, USA

Nearly 50% of high school students in the United States take some form of chemistry before graduation[1]. Typically, chemistry is introduced as a first and only course at the age of 16 or 17 in the third year of high school. This year-long course tends to be long on concept and short on applications. While perhaps appropriate for those students with a strong science orientation, the CHEM STUDY type of approach has been less successful in reaching the "general" student with little prior interest in chemistry.

The public attitude towards chemistry in the United States appears to reflect both ignorance and fear, so that "chemophobia" has become a widespread reaction to the science as it impacts upon society as a technology and in industry. In 1981, the American Chemical Society with financial support from the National Science Foundation began the development of a year-long alternative inter-disciplinary science course called "Chemistry in the Community" (CHEMCOM).

The CHEMCOM approach

CHEMCOM is a course for the citizen concerned with understanding issues at the science/technology/society interface. It is designed for the student who does not intend to become a science major in college. For many students it may be a terminal course, but for others it could be the beginning of a life-long study of chemistry, through the popular media.

Many courses have attempted to make chemistry more "relevant" to this particular group of students — usually by "diluting" the science and adding "appropriate" applications of chemistry. Such courses have begun with the assumption that the structure of the discipline of chemistry, as perceived by professional chemists, is of general interest to the public.

CHEMCOM begins with the assumption that, in general, student interest in certain societal issues involving chemistry precedes that in the discipline itself. Then, the organisation of the curriculum depends on defining that chemistry which students need to know in order to comprehend specific societal issues. In other words, the selection of the issues determines the chemistry to be taught rather than the converse.

The purpose of CHEMCOM, then, is to help students realise the important role chemistry will play in their personal and working lives. This is achieved in part by teaching students how a knowledge of certain principles of chemistry will help them (a) understand many of the technology-related problems they read and hear about in the media, and (b) contribute to solutions of these problems as they become citizens in our "participatory technocracy."

CHEMCOM consists of eight issue-oriented modules each of which takes about one month of teaching time. Each module focuses upon a chemistry-centered technological problem now confronting our society. After identifying the problem, the module explains the chemistry relevant to its understanding and provides information to help students analyse its dimensions. Students also consider solutions to all, or parts, of the problem. This information is assembled by integrating material from the news media with experiments, chemistry theory, and decision-making activities. CHEMCOM provides a "living-textbook" of case histories of unsolved or partially solved problems and of our society's attempts to deal with them. The community context in which these problems exist may be the school, the town, or the region in which the student lives; or the wider world community of "Spaceship Earth."

Topics of the modules are:

Supplying Our Water Needs
Conserving Chemical Resources
Petroleum: To Burn or to Build?
Feeding Our Community
Nuclear Chemistry in Our World
Chemistry, Air and Climate
Our Health and Chemistry
The Chemical Industry: Promise and Challenge

CHEMCOM can be considered a course that examines chemistry through the consequences of its application as technology. In developing the material, great care has been taken to maintain the integrity of the chemistry. As can be seen in Table 1, the eight modules cover the main chemical concepts found in any beginning course.

TABLE 1
Chemical Concepts Utilized in CHEMCOM

Chemical Concepts Introduced or Utilised	Water	Resources	Petroleum	Food	Nuclear	Air	Health	Incustry
Physical and Chemical Properties	I	E	U	U	U	U	U	U
Formula and Equation Writing	I	E	E	U	U	U	U	U
Elements and Compounds	I	E	E	E	E	U	U	U
Nomenclature	I	I	E	U	U	U	U	U
Stoichiometry		I	U	U	U	U		U
Mole Concept		U	E	U	U	U		U
Energy Relationships	I		E	E	E	U	E	U
Atomic Structure					I	U		
Chemical Bonding	I		E	U		U	U	U
Shape of Molecules			I	E		U	U	U
Solids, Liquids, Gases			I	U		U	U	
Reaction Rates/Kinetics					I	U	U	
Acids, Bases and pH	I	I	E			E	U	U
Oxidation-Reduction		U	U	U	U	U		U
Dissociation	I	U		U		U		U
Solutions and Solubility	I	U	U	U	U	U		U
Periodicity		I			U	I		U
Gas Laws and KMT		U	U	U	U	U		
Scale and Order of Magnitude	I	E	U	U	U	U	U	U
Metric Measurement (SI)	I		E	U	U	E	U	U
Equilibrium	I		I			U	U	E
Synthesis			E			U	U	E
Analysis	I	E	E	U		U	U	E

I = Introduced E = Elaborated U = Used

The basic vocabulary of chemisty is introduced in the early modules, as are those fundamental concepts upon which the discipline is structured. Since there is no introductory unit to present this "prerequisite" knowledge the modules must be taught in sequence, as later modules build upon the chemistry introduced previously.

Social science concepts are also presented, again such that students (and the teacher) proceed from the comparatively simple to the more complex. However, CHEMCOM is not a course that straddles that lonely territory between chemistry on the one hand and social science on the other. It is a chemistry course; it will be taught by chemistry teachers; it has been designed by chemistry teachers, revised by chemistry teachers. A detailed evaluation using eight cognitive tests was carried out with about 100 teachers and 2400 students. Some of the items tested the application of decision-making skills developed through the course.

Technology and industry in CHEMCOM

As can be seen in Table 2, the first seven modules all include themes which could be described as "consequences of technological innovation,"

TABLE 2
Technology and Industry Themes in CHEMCOM

Module	"Consequence of Technology	Industries Discussed
Water	Water Pollution	Waste water treatment
Resources	Waste generation	Resources recovery, the recycling centre
	"Lost" metal resources	Mining, ore extraction and refining
Petroleum	Depletion of petroleum resources	Petroleum refining, the petrochemical industry, plastics
Food	Choice of foods	Food processing, agribusiness
Nuclear	Disposal of radioactive wastes	Nuclear power industry
	Exposure to radiation	Radioisotopes in industry agriculture, medicine
Air	Air pollution, acid rain, the greenhouse effect	Smokestack industries, automobile
Health	Improved health and hygiene	Pharmaceuticals

some of them positive, some negative. For example, the Water module is the story of one (imaginary) community's attempt to identify the cause of a fish kill in the local Snake River. During this investigation the citizens of the community (that is, the students) learn a good deal about solution chemistry and the ways in which environmental chemists work. They also examine the technology whereby a municipality purifies its drinking water and treats its sewage effluents.

The Petroleum module investigates our use of petroleum as both a fuel and a chemical feedstock. The central theme is: How can we best utilise petroleum given that resources are limited? Students learn some simple organic chemistry and begin to understand how the petroleum industry operates. The economic importance of petroleum as a feedstock of the organic chemicals industry is stressed, as well as the ubiquity of products derived from petroleum.

A positive examination of the consequences of technological innovation is found in the Food module. Here the food processing industry has provided the students with many nutritional choices. The module is designed to help students make such choices in an informed manner. This unit also expands the student's knowledge of organic chemistry.

In the eighth and final module of CHEMCOM, the chemical industry is viewed as a powerful and sometimes controversial social partner. A company manager introduces a new employee to the organisational and operating divisions of the EKS Nitrogen Products Company. They attend a series of new employee seminars where the new employee learns about the range of career options open within the company.

The module focuses in particular upon the Fertilisers and Explosives Division, permitting an examination of the "two faces" of nitrogen. The chemistry developed includes reversible reactions, chemical equilibrium, redox and acid-base reactions, exothermic and endothermic reactions, reaction kinetics and catalysis, Avogadro's law.

The module also examines the major responsibilities of industrial management; the health and safety of employees and the surrounding community; disposal of waste material; product distribution and economic viability; and, of course, ensuring that all these responsibilities are conducted in compliance with government regulations.

Decision-making component

Each module consists of text, laboratory activities, "Your Turn" problems, and "Decision-makers." The "Your Turn" activities are analogous to the questions found at the end of the chapter in a typical chemistry text. They allow students to develop and extend problem-solving skills including data analysis, graphing, and mathematical manipulations.

CHEMCOM "Decision-makers" fall into three categories:

- Chem Quandary
- You Decide
- Putting It All Together.

The "Chem Quandary" is a short, single-issue problem that students are asked to attempt to solve. There may or not be a single correct answer. For example a "Chem Quandary" in the Water module asks students to find alternatives to standard methods of chlorinating water given that such procedures may form trihalomethanes.

The "You Decide" activities are usually one period in length and are comparable to a chemistry laboratory in that they consist of pre-activity discussions, collecting, and analysing data, making decisions, and post-activity discussion. Often students work on these activities in small, self-directed groups. They may become involved in:

- keeping a diary
- interviewing parents and friends
- library research
- analysing newspaper articles
- proposing solutions to a problem
  in brief class presentations.

A typical "You Decide" in the Resources module asks students to examine the deterioration of the Statue of Liberty and suggest chemically-appropriate way to restore the Statue to its original condition.

Each module culminates in a two-period, "Putting It All Together," where students have an opportunity to use the chemistry they have learned to address the societal concern that is the theme of the module. This may involve the students in role-playing simulations, board games, and debates. The final activity of the Water module, for example, is a simulation of a town council meeting to address the central problem of the module, the fish kill in Snake River. Students decide what caused the fish kill based on the chemical evidence presented and also determine who should be held financially responsible.

References

1. Stake, R. E., and Easley, J. *Case Studies in Science Eduction*, Vol. 1, US Government Printing Office, Washington, D.C., 1979.

26

Chemistry and Industry in Zimbabwe: Providing Resource materials

P. J. TOWSE

University of Zimbabwe

There is a need to provide resource materials for both teachers and students which help to focus attention on issues arising from industrial applications of scientific and technological principles. At present, there is little material available which is appropriate for use in Third World countries.

As a means of providing teachers with resource materials which could be used to generate ideas about science outside the classroom, we selected a wide range of materials from newspapers and magazines available in Harare in Zimbabwe. The series is entitled "Chemistry and Industry in Zimbabwe" and the materials are arranged in six booklets:

Minerals	Resources
Oil	Pollution and the Environment
Natural products	Financial matters

There is a separate introduction for teachers, while each of the booklets contains an introduction for secondary school pupils. These introductions describe how the materials can be used, and each book has questions to stimulate discussion.

Because newspapers and magazines are available in most countries, this example of the local generation of resource materials is one which could be developed by individual teachers, teacher training colleges or science teachers' associations. The collection of materials is relatively cheap and not time consuming, although arrangement of the material under selected headings may take time. Duplication of the materials is not a problem if they are required on an individual school basis, but this becomes an issue if large numbers of copies are required. Revision and updating should be a continuing process as new materials become available.

27

The "Science in Ghanaian Society" Project

J. M. YAKUBU
University of Cape Coast, Ghana

The "Science in Ghanaian Society" Project (SGSP) began in December 1983, supported by grants from Unesco, the British Council and IUPAC Committee on the Teaching of Chemistry. The Project aimed (i) to study local industries in order to identify their underlying science and technology; (ii) to write case studies of those industries; (iii) to develop teaching methods for the case studies; (iv) to relate science teaching to the world of work.

The first phase of the Project was the production of 27 booklets for teachers and an overall guide to the course. Each booklet was concerned with a particular industry. The following were among the case studies prepared for trials:

palm wine industry	food processing industry
vegetable oil extraction	metal working
the soap industry	the brewing industry
the fishing industry	the tanning industry
the salt industry	

For a particular industry, the following aspects were considered.

(1) The geographical location of the village, town, district or region in which the case study was made. The vegetation which was relevant to the industry was described: where a plant is used as a raw material, its biology and ecology was included.

(2) A brief history of the industry in the society of the local people was described. Any legends or folk tales about the origin of the local industry were given.

(3) The various stages of the traditional processes used in the industry were recorded by the case study worker.

(4) The scientific concepts and processes at each stage were identified and a flow diagram was made. The scientific concepts were discussed under each stage.

(5) A flow diagram was drawn of the various stages in the technology of the topic and each aspect was discussed.

(6) The economics of production were discussed with local industrialists. This included the costs of (i) raw materials, (ii) the necessary equipment, (iii) fuel, (iv) labour, (v) overheads. Two tables — the cost of production and the income from sale of the product — were given in order to lead to the idea of profit, which could then be worked out. Each of the industries listed above were discussed in the light of simple economic principles.

(7) The wealth production generated by the enterprise was considered. Attempts were made to show the rate at which the particular industry was growing, the type of people involved in it, how the industry started, what possibilities there are for the employment of school leavers.

(8) The social responsibilities of the industry were considered. What impact was the industry having on society? How does it affect the health of people? Does it lead to pollution of the environment? Does it deplete the natural resources of the environment? These are typical of the questions asked.

(9) Problems of the industry were considered and possible solutions were suggested.

The second phase of the Project began in September 1985 when teachers were introduced to the Project, its content and strategy, prior to comprehensive trials in 50 Ghanaian schools. The teachers were shown how to evaluate the course so that the feedback to the organisers would be as helpful as possible. Following these trials, the teaching materials will be revised and the present intention is to publish the course in 1988.

The target groups for the Project are: junior secondary schools; senior secondary schools; technical schools; teacher training colleges; adult education classes.

Section E

Industrial and Technological Issues: Some Teaching Strategies in the Secondary Science Curricula

Introduction

A tradition of didactic science teaching is strongly entrenched in many parts of the world. The influence of traditional curricula and examinations is so pervasive that many teachers feel constrained to teach only those facts and concepts which appear to be directly relevant to examination success. Yet, if we are to encourage our students to think more widely about some of the issues raised by applications of science in an industrial/technological context, we need to consider a variety of teaching strategies. This in turn raises further problems such as the time available for such strategies, the apparent challenge to the role of the teacher, and the need for preparation of effective resource materials for both teachers and students.

Much of what we are seeking to achieve through the incorporation of industrial issues in the science curriculum is more concerned with attitudinal than with cognitive development. We are more concerned to develop an awareness of the role of science in industry, and industry in society, than to convey mere knowledge of industrial processes. As Reid points out, the normal teaching strategies appropriate for cognitive development are often inappropriate for effecting such attitudinal change.

Traditional, didactic teaching methods (teacher instruction, use of textbooks etc.) are likely to prove ineffective in promoting attitudinal change, and it is suggested that less conventional approaches involving positive participation by the student may be more appropriate, particularly if such activities help the student to move out of the classroom and closer to the realities of industry and technology.

Three teaching approaches were explored in workshops held during the Conference. The first was concerned with the use of simulation games to increase students' awareness of industry and the problems concerned with industrial decision-making. A well constructed simulation can do much to bring the realities of industry into the classroom, and the incorporation of a gaming element ensures active student participation. The paper by Holman is concerned with this teaching strategy. In discussing the use of computer-based games, it was accepted that there were great difficulties caused by the non-standardisation of computer-equipment and the cost of the equipment was felt to make the use of educational computing inappropriate to many developing countries.

Industry is technology in practice, yet much of what currently occurs in science education provides students with little opportunity to experience and practice the processes of technology. The third approach considered is concerned with technological problem-solving activities that can be simply carried out in school. Through the use of such activities, students gain first-hand experience of the processes by which technology is used in everyday life.

However, simulations and problem-solving in class cannot be a wholly satisfactory substitute for reality, and the second approach described in this section is concerned with showing students the reality of industry through the use of industrial visits. Any industry, no matter how small or of what nature, can provide opportunities to see scientific principles in action and industrial practices at work. But the difference between success and failure in an industrial visit lies in careful preparation and structuring by the teacher and the workshops explored way of maximising the potential of visits.

28

Using Games and Simulations to Introduce Industrial and Technological Issues

J. S. HOLMAN

Watford Grammar School, Watford WD1 7JF, UK

A definition of terms and an example of a simulation-game

For the present purposes, a game can be described as a contest between players operating under constraints (rules) in order to achieve an objective (winning or pay-off). A simulation can be regarded as an on-going representation of certain aspects or features of a real-life situation. This paper is concerned with simulation-games, which share features of both, as illustrated below.

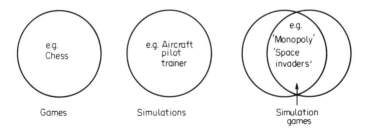

| Games | Simulations | Simulation games |

Simulation-games are commonly used outside the educational field — many popular games, such as the board game Monopoly or electronic games of the space invader type — are simulations, as are war games. Here we are concerned with the educational application of simulation games, particularly those relating to industrial topics.

The Amsyn Problem serves to illustrate the technique. The game simulates a small chemical firm, Amsyn Ltd, in an area of high unemployment. The company manufactures aromatic amines, and during the manufacturing process a certain amount of toxic waste is discharged into the local river, with serious environmental effects. The local

159

government council wants the firm to clean up the pollution, but finding a suitable method is difficult and could be expensive. Failing to find a solution to the problem could lead to the firm having to close down.

After an introduction to the problem, students adopt roles, representing the three major interested parties — management, trades unions and local government. Between them they must consider the various alternative courses of action and arrive at an acceptable solution. Through playing the game students learn a good deal of worthwhile chemistry and, at the same time, develop an awareness of the complexity of a typical industrial problem, and a realisation that solutions are rarely clear-cut. Students practice inter-personal skills of communication, debate and compromise as they experience this simulation of real-life decision-making.

Educational advantages of simulation games

Motivation. Students enjoy simulation-games. They make a change from more normal classroom and laboratory activities, and the element of competition and the need to reach a solution gets students involved to an extent that other activities may fail to achieve. Students enjoy using their initiative and welcome the opportunity to think creatively.

Development of skills. Many of the skills demanded by real-life problems, such as problem-solving and decision-making, communication, co-operation and other interpersonal skills, are difficult to develop through the conventional methods of science education. As already exemplified above, simulation-games provide students with opportunities to exercise and develop such skills.

Insight into real-life problems and situations. Few students, or teachers for that matter, have the opportunity to gain first-hand experience of working in industry, but a good simulation can provide the next best thing. Simulations also make the important point that decisions are rarely clear cut in real-life and usually involve negotiation, compromise and an imperfect solution. This is an important lesson to learn, because students frequently see issues in naive terms and expect perfect solutions. (This may be due at least in part to the fact that many of the problems we present them with in science lessons do indeed have clear cut solutions.)

Interdisciplinarity. Simulations, like their real-life counterparts, generally involve an interdisciplinary approach with, for example, elements of mathematics, geography and economics involved as well as science. They provide a welcome, and all too rare, opportunity to break down subject boundaries, and encourage students to consider problems and issues from other than purely scientific standpoint.

Problems involved in the use of simulation-games

Simplification. Real-life situations are invariably highly complex, and the development of a simulation involves simplification and omission of less important details. The extent to which such simplification and omission are necessary naturally depends on the age and ability of the students for whom the simulation is intended.

Which details should be omitted? Are the simplifications acceptable, or do they distort the truth to an unacceptable extent? The success of the simulation will depend to a considerable extent on the way these problems are tackled.

For example, the Science and Technology in Society Project includes a unit entitled *Dam Problems*. This is a simulation of a procedure used to assess the potential environmental effects of hydroelectric dam projects. A typical environmental impact assessment might consider 16 actions and their impact on 34 environmental categories. In this simulation, the numbers were reduced to 3 and 6 respectively in order to make the simulation simple enough for 15 year-old students to use. By careful selection, it was possible to achieve the necessary simplification yet retain the essential elements of a real-life assessment. As with any simulation, it was important to have the final version checked for accuracy by experts.

Finding time. Simulation-games can be quite time-consuming. The simulation itself is likely to require over an hour of classroom time, and introducing and preparing will require further time. Some teachers may feel they cannot afford such time for what they may regard as a "fringe" activity; yet as has already been emphasised, the benefits in terms of skills and motivation developed can more than justify the time.

The teacher's function. Using simulation-games involves the teacher in a change from his or her usual role. Instead of dispensing facts and theories, the teacher is more likely to become a neutral chairman or even a passive observer. For example, *The Coal Mine Project,* another unit from the Science and Technology in Society Project, asks students to imagine that a coal mine is proposed on a site near their school. Students play roles for and against the project, for example a trade union official and an environmentalist. A student acts as chairman for the simulated Public Inquiry. The only task for the teacher is to ensure that the students are suitably prepared with all the necessary materials, and that the simulation runs properly: he or she does not need to intervene unless things go wrong. This apparently simple task may not be as easy as it sounds, and will certainly involve the teacher in a change of role with which some may find it hard to come to terms.

Examples of interactive curriculum materials

Title	Brief description	Origin and reference	Grades
End to poverty game	A simulation-game which considers the ways a developing country can generate wealth. Calculations and decisions are made concerning labour, technology, education, etc., and effects of natural and social events such as drought and strikes.	Science in Ghanaian Society Project[1]	9–12
Aluminium crisis	Students assume that supplies of aluminium ore to USA are cut off. They divide into groups to decide how to cope with the crisis: aluminium conservation and material substitution; aluminium recycling; aluminium re-use. Groups prepare recommendations and present them and class decides on best course.	CHEMCOM project in USA[2]	10–11
Alternative energy project	Imaginary island off west coast of Scotland has to decide how to meet its future electricity requirements. Groups consider wind power; tidal power; solar power; hydroelectric power; peat. Best use of resources is decided upon.	Science in Society project[3]	11–12
Public Inquiry project	A role-playing simulation related to the siting of a petrochemical plant	Science in Society project[3]	11–12
Buenafortuna Minerals project	A decision-making exercise concerning a fictitious country which has to decide to exploit its coal, uranium, copper or beach sands. Involves scientific, social, economic and environmental issues.	Science in Society project[3]	11–12
The Protein problem	A structured discussion dealing with the world problem of protein shortage. It explores new and traditional sources of protein.	University of Glasgow[4]	9–10
Chemistry in our lives	Students work in small groups to prepare short lectures based on data supplied and on their own knowledge to discover the part played by chemistry in food production, health, transport and in the home.	University of Glasgow[4]	9–10
The Energy problem	Students compete in groups to plan for the development of adequate primary energy for their country for next 50 years.	University of Glasgow[4]	9–10
Energy for the Future	A short programmed learning sequence to enable students to consider problems which will arise when natural gas supplies are exhausted.	University of Glasgow[4]	9–10
Focus on Lead	The unit examines the extraction, purification and uses of lead and provides a social and industrial dimension to the metal's chemistry.	University of Glasgow	9–10

The Amsyn problem	A role-playing simulation involving Amsyn Co. which is faced with closure due to new anti-pollution conditions imposed on it.	University of Glasgow[4]	11–12
Sulphuric Acid story	Students in groups decide where to site a sulphuric acid plant given data from three different periods in history. The divergent conclusions reveal the factors which affect the movement of industry from place to place.	University of Glasgow[4]	9–10
Dam problems	A role-playing simulation concerning the environmental impact of a hydro-electric plant.	Science and Technology in Society[5]	9–10
Coal mine project	A role-playing simulation concerning the social and environmental effects of opening a coal mine.	Science and Technology in Society[5]	9–10
The Limestone Inquiry	A role-playing simulation concerning the quarrying of limestone.	Science and Technology in Society[5]	9–10
Cross-Channel Link	Decision-making exercise concerning the building of a bridge or tunnel across the English Channel.	Science and Technology in Society[5]	9–10
Simcolpa	A laboratory-based simulation of an industrial control laboratory.	Manchester Polytechnic[6]	11–12
The Trading Game	Shows how international trade can benefit and hinder the economic development of different countries.	Christian Aid[7]	11–12+
Uranium	Exploitation of uranium ore reserves in the Central African Republic and the transportation of the ore for processing in France.	University of Zimbabwe[8]	10–12
Ethene	A decision-making exercise concerning the manufacture of ethene from different feedstocks.	University of York[9]	11–12
Ammonia and Protein	Describes the Haber process and students compare the energy efficiency of production of natural and synthetic protein.	University of York[9]	11–12
Chlorine/ sodium hydroxide	A role-playing exercise on the choice of electrolytic cell.	University of York[9].	11–12
Methanol	Balancing the C/H ratio. Selecting feed-stocks for the organic chemical industry in different parts of the world.	University of York[9]	11–12
Mass Spectrometry	Shows an analytical technique in general application. Concentrates on the choice of instrument by "fitness for purpose".	University of York[9]	11–12
Aluminium	Economic factors in location and operation of aluminium smelters.	University of York[9]	11–12

References

1. Science in Ghanaian Society. For further details: J. M. Yakubu, University of Cape Coast, Cape Coast, Ghana.
2. Chemistry in the Community (CHEMCOM). For further details: American Chemical Society, 1155 Sixteenth Street NW, Washington DC 20036, USA.
3. Science in Society Project. For further details: Association for Science Education, College Lane, Hatfield, Herts AL10 9AA, UK.
4. University of Glasgow. For further details: Scottish Council for Educational Technology, 74 Victoria Crescent Road, Dowanhill, Glasgow G12 9JN, UK.
5. Science and Technology in Society. For further details: Association for Science Education, College Lane, Hatfield, Herts AL10 9AA.
6. Manchester Polytechnic. For further details: Dr D. McCormick, Department of Chemistry, Manchester Polytechnic, Manchester M15 6D, UK.
7. Christian Aid Publications, London SW9 8BH, UK.
8. University of Zimbabwe. For further details: Department of Curriculum Studies, University of Zimbabwe, Harare, Zimbabwe.
9. University of York. For further details: The Science Education Group, Department of Chemistry, University of York, Heslington, York YO1 5DD, UK.

Some microcomputer simulations

The following are outline descriptions of some microcomputer simulations, some of which were used in the workshop at the Bangalore conference.

(1) Siting an aluminium plant (from Longman Software, York YO3 7XQ, UK).
This enables students to consider some of the factors which need to be taken into consideration when planning the siting and the production targets of an aluminium smelting plant. The factors which students have to consider are:

a. the source and cost of bauxite,
b. the cost of purification of bauxite,
c. the cost of transporting raw materials,
d. the cost of building a power station and generating electricity,
e. the location and size of the smelter,
f. the repayment of capital cost and interest charges,
g. operating costs,
h. environmental disruption.

Intended for chemistry and other science students aged 14–16 years.

(2) CONTACT, the manufacture of sulphuric acid (from Longman Software, York YO3 7XQ, UK).
This enables students to investigate the effect of reaction conditions (temperature, pressure and catalysts) on the yield of sulphuric acid in the Contact Process. They are able to investigate the economics of the process by varying conditions and investigating the effect of these

variations on the different costs involved (raw material costs, fuel costs, fixed costs, etc). They attempt to find the "ideal" conditions needed to produce sulphuric acid at minimum cost, without causing excessive pollution. Indended for chemistry and other science students aged 14–16 years.

(3) Nuclear reactor simulation (from Longman Software, York YO3 7XQ, UK).

This enables students to consider the effect of changing conditions on the behaviour of a nuclear reactor core and the associated boiler and turbines. A large number of variables can be controlled (for example, cooling water flow rate, position of control rods) and the effect of changing them can be assessed. The behaviour of the reactor core, the boiler and the turbines can be considered separately, or the behaviour of the system as a whole can be investigated. Intended for physics and other science students, aged 14–18 years.

(4) The Paraffin file (from British Petroleum Educational Service, London EC2Y 9BU, UK).

A marketing simulation in which students attempt to find the best combination of market variables for the sale of paraffin (kerosene). Variables that can be controlled include price, advertising, sales staff, and market research. At the end of each "year" of sales, students are given their sales results and told their share of the paraffin market. Intended for economics and business studies students aged 16 upwards.

(5) MINOS (from Microelectronics Education Programme, Leeds Polytechnic, Leeds LS1 3HE, UK).

A simulation of the MINOS computer system used in British Coal Mines. The user is shown a plan view of part of a coal mine and is given control of the shearer for cutting coal, a drill for making a tunnel, conveyors for carrying the coal, and an air duct and fan. The objective is to cut coal at minimum cost, taking into account the need to keep within safe limits of methane (firedamp) build-up, and to carry out repairs and maintenance.

(6) Ethene (from Science Education Group, Department of Chemistry, University of York, York YO1 5DD, UK).

A set of seven programmes. Students can investigate costs of four possible feedstocks, investigate the optimum conditions for operating cracker furnaces and study the effect of economies of size or production costs. The final programme is a mangement game in which students attempt to match output and demand.

(7) Refinery (from Project Seraphim, Department of Chemistry, Eastern Michigan University, Ypsilanti, Michigan 48197, USA).

This comprises a tutorial which explains the various operations of a modern refinery and a game in which the player who is the operations manager of a refinery makes decisions about operating the refinery in order to maximise profits.

(8) Sulphuric acid (from Project Seraphim, Department of Chemistry, Eastern Michigan University, Ypsilanti, Michigan 48197, USA).

This comprises a tutorial on the manufacture of sulphuric acid and the game allows the user to experiment with reaction conditions to obtain the best economics of production.

(9) Octane (from Project Seraphim, Department of Chemistry, Eastern Michigan University, Ypsilanti, Michigan 48197, USA).

Tutorial and game on gasoline, octane rating, engine knock and compression ratios and gasoline additives. The game involves a simulated drive for which gasoline must be blended to match the characteristics of the car.

Commentary after the workshop on simulation-games

In general, there was enthusiasm for the technique of simulation-games which was new to many of the participants. It was felt that they were of value in developing useful societal skills; and that their interdisciplinary nature was a good way of effecting links both between subjects and within a particular subject. However, it was acknowledged that they are time-consuming and that debriefing following the game would be essential.

Participants felt that this technique could lead to a difficult change of role for the teacher, particularly in cultures where there is a tradition of didactic teaching. In this respect it would be helpful to provide the teacher with as much background information as possible.

There was discussion of whether the term "game" was appropriate in this context. There was a feeling that this term tended to trivialise what could often involve a serious and profound discussion.

There was further discussion of the role of science in the games that had been played. In several examples, the science component was small and some participants wondered whether, in view of this, the time spent playing the games was justified. Others felt that the advantages gained from the development of more general skills certainly justified their use.

There was also discussion of the applicability of the technique of simulation-gaming to cultures other than those in which they had been developed. In general, it was felt that the technique itself was indeed generally applicable, though naturally a game designed specifically for use in one culture would be of little direct use in a different culture. "Adapt, not adopt" was felt to be an appropriate maxim: the technique should be adapted to suit local conditions.

29

A Laboratory-Based Industrial Simulation — An Experiment

D. McCORMICK

Manchester Polytechnic, UK

For those considering the introduction of games and simulations into science courses, there might be some interest in a trial industrial simulation which we carried out in the Chemistry Department at Manchester Polytechnic a few years ago. The Technical Education Council, the examining body for one of our sub-degree courses, had decreed that a component of the course should be an industrial experience. They suggested that a "less than ideal" alternative to this would be a college-based industrial simulation. Since neither they nor we knew much about such simulations, we decided to run one on a trial basis. We entered into discussions with the Colgate-Palmolive Company, and took up one of their suggestions to plan a simulation of the activities of their control laboratory in Salford. The simulation was planned in great detail. The students would work a normal working-day, carrying out routine analysis on the Company's raw materials and products. They would also have to respond to requests for repeat determinations, and to various crisis situations which our Colgate-Palmolive collaborators had written into the programme in the form of instructions held in sealed envelopes which were to be opened at specific times. The students were to run the laboratory by themselves, their only contact with us being by telephone to the simulation co-ordinator. We ran the simulation, known as SimColpa, with eight student volunteers who were given a nominal wage, which was provided out of funds from the Department of Industry. The simulation lasted two weeks, and was a salutary experience for all concerned.

After an initial 2-day familiarisation period, the simulation proper started. For the first few days of this, there was frenzied activity, and the students were physically exhausted. Slowly but surely, they started to master their work, and to organise themselves with far greater efficiency. Those who had to act as laboratory supervisor took their work very seriously indeed, even skipping coffee breaks to catch up. There were occasions during the simulation when it became all too real: a crisis situation resulted in a very serious argument between two students! During the two weeks we sounded out student opinions and attitudes by feedback sessions and by questionnaires. These convinced us that with regard to objectives such as encouraging collaboration, initiative, and the critical interpretation of observations, the simulation was far superior to normal laboratory instruction. The same could be said of the way in which the simulation emphasised the importance of organisation, accuracy, and mastery of technique. It was only in the area of emphasising the underlying chemistry principles that normal laboratory teaching situations proved superior.

There is also one comment which needs to be made: that towards the end of the two weeks, there were undoubted signs of boredom creeping into the students' behaviour.

Since running the simulation trial we have been able to place most of our students in industry. On the one occasion when we were unable to do this, we ran a SimColpa[1] exercise with many of the same results as mentioned earlier. We prefer to place our students in industry since there is really no substitute for that experience. However, we have not forgotten the lessons learned from the simulation. Chief amongst these is that normal laboratory instruction has many shortcomings, in both its general and vocational functions.

Reference

1. SimColpa: An Investigation into Laboratory-based Industrial Simulation. M. P. Coward, D. B. Hobson, D. McCormick, J. G. Quinn and D. A. Tinsley. *Perspectives in Academic Gaming and Simulation* **6**, pp. 189 to 200, Kogan Page: London 1981.

30

Interactive Packages for Teaching Industrial Issues

N. REID

Craigmount High School, Edinburgh, UK

In Scotland, pupils may take the Ordinary Grade Examination in Chemistry at age 15–16, followed by the Higher Grade Examination one year later. The Higher Grade Examination is the qualifying examination for entry to Higher Education. Pupils can also choose to stay on for one further year at school and may take a further Chemistry Examination known as the Sixth Year Certificate. Chemistry is a popular subject and the proportion of pupils choosing to take Chemistry at school has risen steadily for at least two decades. Indeed, at the present time, Chemistry is the second most popular subject at the Sixth Year Certificate.

The examination syllabuses at all three levels emphasise the social, industry and economic implications of chemistry. It has long been thought that pupils should be able to see chemistry as it is applied in life both at local and national levels. However, in the period up to 1975, there was no sustained attempt to provide teaching materials specifically designed to meet these social, industrial and economical aims.

Knowledge or attitudes?

One of the major problems has been a lack of clarification in the aims of the syllabus. It is too easy to interpret the industrial aspects of a syllabus in terms of the knowledge given to pupils. It has to be remembered that such industrial knowledge rapidly dates and, for most pupils, is a complete irrelevance. Around 1975, it became clear that the important aims relating to the industrial aspects of chemistry were attitudinal in nature. This was a large breakthrough because it allowed the construction of teaching materials that were designed to bring about attitudinal growth rather than cognitive growth.

It is important to define the kinds of attitudes involved. It was never intended that pupils should be manipulated in some way to accept certain prescribed viewpoints. The aim was to allow pupils to develop their own attitudes based on informed awareness. For example, in considering chemical industry, pupils were not to be taught the facts of various chemical processes; neither were they to be given some view of industry that was thought to be "desirable". At the end of the course, they should have an informed appreciation of why the industry exists, where it fits into local and national economies, why it develops in certain geographical locations, why processes often have to change and particular production units have to close while others open, and so on.

Most of the educational output from the chemical industry seems to be aimed at achieving straightforward cognitive aims. Most textbooks of chemistry aim to communicate the facts (often out of date) about key industries. The great lack, in 1975, was for teaching materials that worked THROUGH industrial chemistry to give pupils an appreciation of the industry and how it operates, without losing the pupils in the vast array of industrial data.

To meet this perceived lack, the Scottish Education Department funded a large project with the intention of providing such teaching materials and studying whether the materials produced actually did achieve the intended attitudinal growth in school pupils. At the outset, the Ordinary Grade course (ages 14–16) was considered. Later, material for other courses became available.

Attitudinal growth

There are many well established methods by which cognitive growth can be achieved. However, it cannot be assumed that attitudinal growth will arise simply by using strategies that have proved useful for achieving cognitive aims. In order to consider the possible ways by which attitude growth might occur in classroom situations, it is necessary to review carefully the findings of social psychologists working in the field of attitude change.[1]

In studying their results, it becomes clear that, except for experiments in psychological laboratories, attitude change tends to occur gradually, it seems to need to be "controlled" in some way by the subject and it often involves deep levels of personal involvement. Clearly, didactic type approaches, — whether of the "chalk and talk", the glossy magazine or the well presented film, — do not seem to be effective in attitude development.

In reviewing the literature of social psychology and related areas, the suggestion was made in 1976 that, in order to allow for attitude growth, a process of interactivity was involved. In interactivity, the pupil becomes deeply involved with the subject matter in such a way that cognitive or affective input interacts mentally with previously held attitudes and beliefs. If this hypothesis is true, then there is a need for teaching materials likely to bring about this internal interactivity. These materials became known as "interactive packages". It should be noted that this use of the word "interactive" has nothing whatsoever to do with the study of interactions that can occur in classroom situtions. Interactivity is a postulated internal process which may or may not involve overt interactions.[2]

Strategies for interactivity

One of the most powerful tools for achieving attitude change as noted in the literature is the use of role play.[3] In role play, pupils may well have to interact with facts and viewpoints that are not compatible with their own attitude corpus. The extent and pace of involvement is, to some extent, controlled by the pupil in that he may be able to limit his identification with the role being played.

In a similar way, constructing or delivering lecturettes can provide interactivity: material has to be studied, mastered and re-structured in a way suitable for presentation. Essay writing can achieve the same depth of involvement but it is easier for the pupil merely to present the ideas to others without real involvement.

Very often, discussion in groups achieves very little. Discussions can "wander" aimlessly; sometimes discussions do not involve all the potential participants; very often, participants do not actually become involved with the input from others in the group. However, it is possible to structure a discussion in such a way that all the group have to participate and it is possible to design the discussion in such a way that the participants have to interact with each other and with the material under consideration. One way of doing this is to form a group of, say, six. Each of five participants are provided with separate, but different, sheets of information; the sixth member is given a sheet of questions.

After a few minutes for reading, the sixth member starts the discussion by asking the questions. The first few questions are simple and require factual answers. Only one member of the group has each answer — so everyone is "forced" to contribute in a simple way. Later questions require that information from two or more members of the groups "is" brought together to give an answer. The final questions require information from all five sheets and the whole group is by now totally involved. This kind of structured discussion has been filmed and the film illustrates a very neat example of the structured discussion group being an effective method by which attitude growth takes place.

There are a number of other strategies that seem to be able to bring about interactivity: these include decision-taking exercises. Examples are available which involve co-operation and others which involve competition. Competitive exercises also seem to be interactive in nature. Various simulation techniques can be adapted to bring about interactivity.

The interactive package

When constructing interactive packages, it is important that the structure of each package is designed very carefully in order to bring about maximum interactivity. However, several other features are vitally important if the package is to be used widely in chemistry courses.

The subject matter of the package must be directly related to the appropriate chemistry syllabus. If the content is not perceived to be relevant to the syllabus (and thus to examination success), the teacher will not use the package. Secondly, the package must be reasonably short. Ideally, it should fit comfortably into the normal unit of teaching time, which, in Scotland, is frequently around 80 minutes. Many of the packages are much shorter than this and can be integrated into other activities, making the packages a natural part of the teaching flow. Thirdly, the packages must be self contained. Teachers should not have to search for information, experimental details or further instructions. The package should have a brief but clear teacher's guide which sets out the aims of the package and gives simple instructions as to its application. Finally, the packages must be produced in a way that makes them cost effective. Ideally, teachers should be able to invest a small amount of money in purchasing the packages which can then be re-used over several years.

It has become clear, through extensive trials as well as from large scale sales, that these practical features of the packages are far more important to teachers than any postulated theories about interactivity! Fundamentally, teachers want materials that work well in class and which are cost effective in terms of both time and money.

Same content — different format

Almost all teachers complain from time to time about the overloading of the syllabus. In Scotland, chemistry teachers are extremely fortunate in that the syllabuses tend to be reasonably short. However, teachers will not willingly bring in extra work into an examination syllabus.

The interactive package idea was designed to avoid this problem. The packages cover traditional chemistry content, thus adding little or no extra burden on to the syllabus. What they do is to present familiar chemistry in an entirely new way, a way which brings about interactivity. In this way, the intention was to teach familiar chemistry in such a way that there was the added bonus of attitudinal growth.

For example, it is possible to teach the industrial side of sulphuric chemistry by the traditional didactic presentations, illustrated by appropriate laboratory work. In the interactive format, pupils work in groups. Each group is provided with the information from a different point in history. Each group has to plan a factory to produce the acid under these historical conditions. This may take only 40 minutes. The groups then compare answers. To their surprise, they find that their answers differ. They then can perceive that industrial decision taking is not a fixed static process. They can see that decisions vary with time and processes can become obsolete. They also grasp readily the importance of the uses for a product in influencing its manufacture. Appropriate practical work can now follow.

The key fact is that they have learned the same chemistry as they would have done through a traditional presentation. However, with NO extra time demand, they have interacted with the information. It is also observed that the pupils have enjoyed the whole process in a way rarely observed in more traditional approaches.

Many topics have been covered in the interactive packages.[4,5] Traditional topics like the manufacture of sulphuric acid, the development of the Haber Process, and applications of isotopes are available. In one package, the world protein problem is looked at from a chemical standpoint, while, in another package, pupils compete to manage the development of primary energy resources over a 50-year period.

One package which is popular with pupils involves the uses of synthetic and natural fibres. It is usual to discuss polymerisation, properties of polymers and then their uses. In the package, this order is reversed: the market demand requires fibres with specific properties and pupils have to find the correct polymer to match these requirements; a look at structure follows. This is typical of many of the packages, where familiar chemistry is "turned on its head" in order to provide an interactive situation.

For older pupils, the scope for devising such packages is much greater. In one such package, pupils have to role play in dealing with a simulated pollution problem that is affecting a complete community. Most of the original set of packages were written in 1976–7.[6]

In 1981, a second set was published as an integral part of a chemistry textbook.[7] In 1982, another set of six packages was written to meet the needs of the Sixth Year Chemistry syllabus. These followed the well established interactive structure.[8]   .

Evaluation

Most of the packages that were written in 1976–7 underwent extensive evaluation. This involved 1100 pupils in typical Scottish schools. The evaluation was in two phases. In the first phase, each package was checked to see if it worked in action with pupils. This involved checking the language used, the clarity of instructions, the time taken, the accuracy of this data etc. The second phase, however, was much more important. This attempted to find out if the packages, over the long term, actually did achieve the attitudinal objectives that had been specified.

From the vast amount of data gathered, it became clear that many of the attitudinal objectives were being achieved, often to a surprising degree. However, it was also clear that there were clear limits in the type of objectives that could be achieved: for example, it seems that it is quite impossible to use the interactive technique as a "brainwashing" instrument Personal attitudes, it appears, are not susceptible to influence through this kind of interactivity.[9]

Conclusions

The interactive package was developed to meet a real need in chemical education. Evaluation indicates that there is optimism that such packages are useful as a method for achieving objectives that are attitudinal in character. Other research has demonstrated that didactic techniques are not successful in this area. The concept of interactivity has been put forward as a possible mechanism by which some attitudes can develop and there has been some discussion about the mechanisms of this process. The interactive packages in chemistry are offered as another resource to assist the teacher in making the study of chemistry an enriching and rewarding experience.[10]

References

1. Reid, N. *Attitude Growth and Measurement — Review*, Report, Scottish Education Department, 1975.
2. Johnstone, A. H. and Reid, N. *New Materials for Chemistry Teaching*, Bulletin 15, Scottish Curriculum Development Service, Dundee Centre, **15**, 2–29, 1979.
3. Reid, N. *Simulation Techniques in Secondary Education: Affective Outcomes*, Simulation and Games, Sage Publications Inc.: California, 1980.
4. Reid, N. *Bringing Chemical Industry into the Classroom*, Chemistry and Industry, February 17th 1979.
5. Reid, N. *Chemistry with a Time Dimension*, Education in Chemistry, **19** (6), 1982, 166–168.
6. Reid, N. *Interactive Chemistry Packages*, Scottish Council for Education Technology, Dowanhill, Victoria Crescent Road, Glasgow, Scotland, 1978.
7. Johnstone, A. H., Morrison, T. I. and Reid, N. *Chemistry About Us*, Heinemann: London, 1981.
8. *Royal Society of Chemistry Sixth Year Units*, Scottish Council for Educational Technology, Dowanhill, Victoria Crescent Road, Glasgow, UK, 1982.
9. Reid, N. *Attitude Development Through a Science Curriculum*, Ph.D. Thesis, University of Glasgow.
10. Johnstone, A. H. and Reid, N. Towards a Model for Attitude Change, *European J. of Sc. Ed.*, **3** (2), 1981, 205–212.

31

Encouraging Student Participation in the Broader Issues of Science

C. J. GARRATT and B. J. H. MATTINSON
University of York, UK

Science teachers have to be selective in what they teach. The subject is now so vast that it is hard to reach agreement over even the range of basic facts to be included in a school syllabus. Furthermore, facts themselves are of limited value without an appreciation of how they might be used in a social context. There is much debate about the consequent need to shift the emphasis in science teaching from the assimilation of information towards the development of skills needed to use that information. These skills, like all practical skills, must be learned through experience rather than wholly by didactic teaching. They differ, however, from skills learned through conventional laboratory work and, to practise them, students must experience new situations. Case studies provide a useful means of generating such learning skills.

Three such studies are used by undergraduates reading Chemistry at the University of York, as part of an option which concentrates on the particular contribution which the application of chemical principles and knowledge can bring to the problem of reconciling the exploitation of natural resources with protection of the environment. These studies, which take three or four sessions of two and a half hours, deal with the forecasting of demand for fuels, the development of a technical strategy for exploiting an oil field in the North Sea, and the planning of a new coal mine in the light of social and economic constraints. Uncertainty clouds the issues, and judgements require that the limited facts be tempered by social, economic and environmental considerations. Many students find it difficult to come to terms with this loss of the precision which they associate with science and to which they are accustomed. The studies are designed to help them to overcome these difficulties.

Early experience with case studies

All the case studies examine a real or realistic situation and involve some degree of student participation and judgement. Beyond these common features, attempts at a succinct, comprehensive definition fail because the term is used to cover such a wide range of materials available to teachers. The purpose of using these materials is not always clearly defined, beyond an implied expectation that they will stimulate student interest. If the approach is to achieve its potential, it is important to analyse its advantages and limitations.

The York studies for undergraduates were designed to encourage the practical use of information and not simply to teach information itself, which could probably be done didactically in a quarter of the time. The participants are required to use scientific and social criteria to make judgements and decisions based on data which are necessarily incomplete and imprecise.

For example, in considering the development of an oil field in the North Sea, students use apparently simple information as a basis for making judgements about the placing of deep sea production platforms and thus begin to appreciate the uncertainties of the data.

School teachers have long advocated that the skills which these studies are designed to develop should be a normal part of science education. We therefore decided to adapt the first of the York Studies for use by sixth forms. This package examines the patterns of energy use in UK households which seemed a suitable subject for general studies programmes: it is within the everyday experience of schoolchildren; the information content is neither very difficult nor technical; all responsible householders must concern themselves to some extent with their domestic fuel supply. We hoped that the interest of the pupils would be stimulated not only by the obvious relevance of the subject but also by the emphasis on "doing" and "discovering". The recast study was tested in the sixth forms of local co-educational schools. The classes varied in size from 12 to 20 and, within a fairly narrow age range (16–18 years), there was a wide range of ability and commitment.

These early experiences in the University and in schools led to general conclusions which, if not original, are not always stressed by enthusiasts of case studies.

(1) Relevance is subjective. In order to stimulate interest, it must be recognised not only by the teacher, but also from the different perspectives of the student who is a young person first, a future adult second. For example, our expectation that the study of domestic energy usage would be seen as relevant to 16–18-year-olds was often incorrect.

(2) "Learning by doing" is not necessarily the best, or even a good way to learn facts; it is slow and may be confusing. Nor is "discovery learning" necessarily stimulating if there is obviously a single correct answer which the teacher knows from the start. Committed student participation develops out of interest: interest does not necessarily develop from enforced participation.

(3) "Learning by doing" is a good way to learn skills including the comprehension of written material, the interpretation and use of data, the application of elementary mathematics, communication, co-operation (working in groups) and rational decision-making in the light of conflicting information.

These conclusions have strongly influenced the structure and organisation of the studies which have subsequently been revised or initiated in this department.

The design of the York packages

Growing awareness of the problems of preparing effective and practical teaching material of the kind described, and of the increasing demand for it, led to the formal identification in 1983 of the University of York Science Education Group. The group has aimed to produce packages of material which are self-contained and ready for use by the teacher with a minimum of preparation.

The group consists of a loose association of teachers with a permanent core of University staff in the departments of Chemistry and of Education. This association provides a network of a wide variety of schools and teachers giving ready access to a growing pool of experience and to the test-beds necessary to develop material for general publication.

The group has also built up a wide range of contacts in industry and commerce. These play an important role in providing both background knowledge and detailed information to individual members of the group preparing packages of teaching material. Preliminary drafts of these are criticised by other teachers before an initial test in schools; revised drafts are used by teachers in their own schools to provide a realistic trial. Publicaton of packages began in 1984.

In planning the packages we have found it helpful to consider content and presentation separately.

Package content

The subject-matter of the study and the skills which will be practised must be clearly evident to the teachers, who must also be able to recognise that use of the package will help in the achievement of their educational aims.

The choice of subject is governed by the relative emphasis to be placed on the acquisition of scientific information, on social and economic considerations and on the practice of skills. This balance will differ for different studies. The York packages deal with topics in Chemistry syllabuses building on examinable scientific information to explore social and economic considerations. At the other extreme, the practising of the skills of, for example, assessing the economic and social case for a particular type of development might have a greater impact if the information vehicle were selected for its immediate relevance as perceived by the young student: proposals to construct some nationally significant chemical plant or power station may be of insufficient concern to students to stimulate their interest in assessing its social and economic consequences; the principles might be more effectively illustrated by consideration of a local sports facility or supermarket.

The form of student involvement is planned in the light of the chosen balance between information and skills. The York packages include examples of class discussion based on open-ended questions, of the use of work sheets involving the processing and interpretation of data, and of role play. A key element is the opportunity for the student to exercise judgement rather than seek a predetermined answer. Particular care is needed with the planning of calculations since these in excess can lead to loss of student interest.

Package presentation

The package must not be too expensive, require too much preparation nor be too time-consuming. Furthermore, insofar as the teaching approach may differ from the user's normal teaching style, guidance for the use of the material must be clearly given. Particular importance is attached to minimising teachers' preparation, which represents a large investment of time. Each package includes a comprehensive Teacher's Guide which covers the following points:

(1) objectives, with key teaching points at each stage;
(2) the type of student for whom the study is suitable and the extent of prior knowledge required;
(3) a suggested teaching structure with estimates of the time required.

The teaching structure aims to chart a pathway between encouraging open-ended exploration of particular points and giving a closely defined lesson-plan.

In addition to these key points, each package contains the necessary teaching materials: up-to-date information with sources and details of supporting material such as publications, computer programmes, video tapes; master copies of student work-sheets, information sheets, role-briefing sheets and overhead projector transparencies, all free of copyright; suggested calculations, questions and discussion points with answers. Some packages include microcomputer software designed for teacher demonstration, small group activity and individual learning.

Illustrations from the York packages

Each undergraduate package is designed to occupy four periods of three hours. The Chemistry packages for 16–18-year-olds require about one double period plus homework to teach a subject which might be covered in a single period by conventional teaching. The general studies energy package can be used very flexibly to occupy from four to twelve periods plus optional homework depending on the ground covered.

Some illustrations of the approaches used are given below.

(1) In ENERGY IN THE HOME data for 1960 and 1970 are used to forecast the domestic demand for fuels in 1980. This exercise reflects a real activity in which the answer is not known at first and provides an opportunity to learn from discrepancies between student forecasts and the actual figures when revealed.

Earlier in the same exercise, useful energy obtained by households is calculated from fuel supply statistics. Students require estimates of the efficiency of fuel use, which cannot be known with precison; they learn that, in spite of this, useful conclusions can be drawn.

(2) AMMONIA AND PROTEIN involves the quantification of some consequences of converting ammonia to protein either *via* fertilizer and crops or direct as single-cell protein. The calculations show that there is no preferred route, but provide a basis for choice in given circumstances.

(3) Several packages compare chemical processes and, in addition to teaching process-information, illustrate the nature of the judgements necessary when social, economic and environmental considerations are pulling in different directions. For example, in ALUMINIUM, students identify which of several real plants should be closed and compare their judgement with the actual decision.

(4) In THE PRODUCTION OF CHLORINE AND SODIUM HYDROXIDE, role-playing has proved extremely effective in illustrating the conflicts associated with process selection. Presentation of a case to a critical audience provides a severe test of understanding.

Conclusion

The York packages are designed for use in secondary schools and universities in the UK and are unlikely to be suitable for direct use in other situations; relevance alone would be a major problem. However, the general conclusions drawn from them, and the principles used in writing them should be useful in any situation. Intending writers of packages should first define carefully the local need for them and then select suitable topics in the light of dominant local activities and interests.

Appendix

Further details can be obtained from University of York Science Education Group, Department of Chemistry, York University, York YO1 5DD, UK. The University packages include:

(a)   *Oil Strike:* estimating the recoverable reserves in an oil field and planning a strategy for their recovery in the light of technological and economic constraints;
(b)   *Avon Coal:* planning the development of a new coal mine in the light of economic and social constraints.

The Chemistry packages for 16–18-year-olds include:

(a)   *Ammonia and Protein:* the growing demand for ammonia, the Haber process and the effect of reaction conditions on the position of equilibrium, comparison of the use of ammonia as a fertiliser with its use as a feedstock for single-cell protein;
(b)   *Chlorine and Sodium Hydroxide:* a role-playing exercise in which groups of students present the case for manufacturing chlorine using different technologies, emphasising social, economic and techno- logical factors in making decisions;
(c)   *Aluminium:* the production of aluminium, and the associated energy costs leading to discussion of the problem of selecting one (out of five) aluminium smelting plants for closure;
(d)   *Methanol and Synthesis Gas. Ethanoic acid.* This pair of packages covers the manufacture of important organic chemicals and illustrates the pace and effect on the chemical industry of technological change.
(e)   *Mass Spectrometry:* an exploration of the nature and usefulness of the mass spectrometer, particularly in industry.
(f)   *Ethene:* an exploration of the ways in which economic, technological and social factors influence the choice of raw materials for ethene synthesis and the route of manufacture.

In addition the Group has developed the following microcomputer software:

(a)   *Interpretation of Mass Spectra:* programmes simulating data obtain- able from the mass spectrometer and dealing with the interpretation of the data.
(b)   *Economic Aspects of the Chemical Industry:* programmes handling cost data for selected chemical processes (or allowing input of the user's own data) and allowing the study of economic optimisation of plant design and plant operation.

32

Technology as Problem-Solving

R. T. ALLSOP

University of Oxford, UK

The notion of technology as a problem-solving process has been referred to repeatedly in this book. For many teachers any attempt to respond to the acknowledged importance of this process within their own lessons may be seen to be accompanied by very real difficulties. For other teachers problem-solving allows them the opportunity to break free from the straitjacket of routine note-taking or repetitive practical exercises. In order to remind the reader of a possible framework for problem-solving activities, the "problem-solving chain" used by the Assessment of Performance Unit in the UK is given below (source: APU — *Science in Schools Age 15 No. 2* — Department of Education and Science, London, 1984, page 82). Reproduced with the permission of the controller of Her Majesty's Stationery Office.

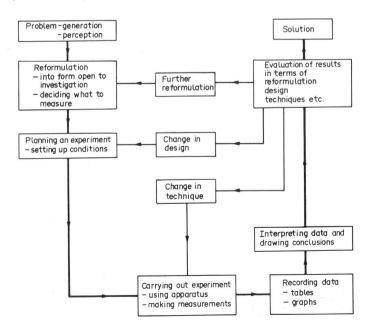

For the teacher who wishes to engage with technological problem-solving activities, a number of possible approaches are available:

(a) Round off a topic in the curriculum by spending one or two lessons on a problem derived from the theme being considered. For example, at the end of a topic on energy conversion with a 12/13-year-old class, ask each working group to construct a device to lift a given mass using water power.

(b) It may be valuable to extend such problem-solving projects over a longer period of several lessons, by suspension of the normal lesson-by-lesson development of the curriculum. This will allow the creation of more elaborate projects with a larger constructional element and also likely to require drawing skills.

(c) Problem-solving projects may be described in such a way as to introduce a *competitive element* into the work. Student motivation is frequently increased by this strategy, for example:

> "Design and build a bridge, from newspaper and a limited amount of sellotape, in order to span a 30 cm gap. Your bridge will be tested to see what maximum load it will carry"

or

> "Design and build a rubber-band powered vehicle to travel 20 m as quickly as possible".

Some examples of the briefs for successful projects are given below:

1. Investigate the stability of different shaped cups.
2. Investigate which is the most absorbent material for mopping up — cloth, kitchen towel, paper towel, tissues etc.
3. Investigate the effectiveness of different floor coverings for such things as quietness, wear, slip, easiness to clean, etc.
4. Investigate the strength of different plastic carrier bags for strength and ease of carrying, i.e. do not cut into a person's hands.
5. Investigate the best way of preventing rust on say four similar pieces of steel.
6. Investigate the road safety of different colours for clothing.
7. Investigate ways of harnessing balloon power, e.g. balloon powered machine.
8. Investigate on which surfaces snails move more quickly, e.g. glass, ceiling tile, floor tile.
9. Investigate whether the height different students can jump is related to leg length, height etc.
10. Devise and make an instrument for measuring the wind or draughts.

Teachers who have used this approach identify many positive outcomes, among them being significant gains in *motivation* and *attitudes to science/ technology*. They particularly report success with students who do not do well in more conventional curriculum activities. As one head of a science department said: "There is clear evidence of those pupils who have not generally succeeded in the more structured side of the science course doing surprisingly well in the technology project".

They also report difficulties or doubts related variously to their perceptions of what students should be doing/achieving in their lessons and to practical difficulties of organisation. The following are common worries stated by teachers:

(i)   Identification of problems that can be undertaken by a large number of pupils, e.g. a whole year group.

(ii) The resulting change in the role of the teacher, away from being the provider of information.

(iii) Reluctance to give such problems to pupils of all abilities, thinking that they perhaps suit only the less able.

(iv) The difficulty of identifying tasks with a biological or chemical flavour, as opposed to the predominant physical emphasis.

(v)   Some see these activities as suitable for science clubs or for homework but not as centrally placed in the curriculum.

(vi) The need to introduce pupils to the use of sophisticated scientific equipment is seen by some teachers to be contradicted by the use of improvised apparatus in problem-solving projects.

The workshop at the Bangalore Conference

Participants were given an opportunity to engage in a problem-solving activity. They were given the following task card and invited to work in groups of about 5 or 6 persons.

TASK CARD

Design and make a device to project an egg launched from the balcony of the room in which you are working (on the *eighth* floor) so that it will land intact at ground level, using *only* the specific materials provided. You have 40 minutes to prepare your device for testing.

Materials: 1 newspaper, 1 m sticky tape, 2 twigs, 4 paper clips, 6 drinking straws, 1 egg.

Throughout the activity, all the participants were highly motivated, involved and enthusiastic. Four of the six eggs landed safely in good condition!

In subsequent discussions, it was suggested that, in an industrial context, the term "problem-solving" had negative connotations, being equated with difficulties in plant operation ("trouble-shooting"). For school usage, the terms "challenge" or "innovation" were thought more positive.

Although this type of work was already frequently seen as a component of successful primary school science teaching, it was accepted that teachers and parents often needed to be persuaded of the value of such activities at the secondary level. It was necessary to convince them that such activities do not take away valuable time from completing the syllabus, but add to the students' understanding and appreciation of core material. Teacher training courses should introduce trainee teachers to the possibilities of incorporating problem-solving activities into their lessons.

The following references provide some useful starting points for those wishing to try out simple problem-solving activities.

References

Philpot, A. and Sellwood, P. "An introduction to problem-solving activities", *School Science Review*. Vol. 65, No. 230, September 1983, pp. 19–32.

Ideas for egg races, British Association for the Advancement of Science, 23 Savile Row, London, W1X 1AB. (A compendium of ideas for competitive projects.)

33

Industrial Visits

J. B. HOLBROOK

University of Hong Kong

Few secondary or primary school teachers have direct experience of industry. This does not imply that they cannot raise and discuss issues of relevance to industry and technology in their teaching. If they provide their students with direct experience of selected aspects of industry, this can help enrich teaching about the applications of science. Industrial visits can form a useful part of an overall teaching strategy in which knowledge and understanding of industry and technology are developed.

Exposure to selected industries can help students to understand that industry and technology are human enterprises with all the potential strengths and weaknesses of such endeavours. It is also possible for students to begin to see that industries often use processes in which scientific principles are applied, as well as a range of technological skills to meet specific needs. The distinction between science and technology may be difficult for students at various levels to appreciate, but we should encourage discussion of this issue (see, for example, Wynne Harlen's paper, page 5).

The economics of industries are often of great interest too. People are employed and have to be paid, raw materials have to be purchased and converted into finished products, and these products have to be sold in order to generate profits, which are needed for further investment. Students may find these issues particularly interesting in a local context when relatives and friends are involved.

Industrial visits may add significantly to students' perceptions of the real world. If such visits are to be useful, the teachers involved must ensure thorough preparation. Without adequate planning and preparation, industrial visits will tend to involve visions of "pipes of varying diameter". The intended outcomes of the visit must be clear to all concerned and the various stages of planning and preparation require considerable work on the part of the teachers. Communication between the school and the industry at every stage is of critical importance. Such a scheme as is described in the next chapter (page 191) can be very appropriate in countries which are highly industrialised. However, for those countries with fewer industries an elegant alternative is suggested in the chapter after that (page 195).

A case study: a visit to industries in Bangalore

At the Bangalore conference, participants were prepared for visits to three industries close to Bangalore. The "Heinola model" (so called from a project in Finland) of an industrial visit (see below) was used as a basis for discussion.[1] This discussion was followed by a video presentation (specially prepared for the Conference) of the essential processes involved in each industry. The participants were divided into groups, each group being asked to prepare a case study of a visit to one of the industries. The model for the pupils' visit to industry was as follows:

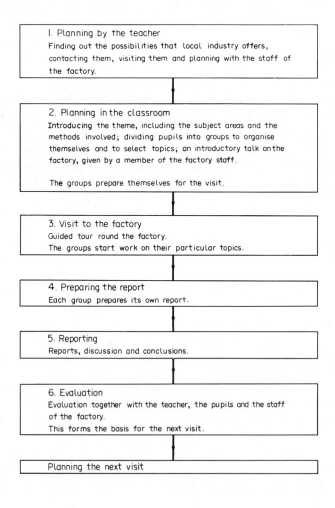

1. Planning by the teacher
Finding out the possibilities that local industry offers, contacting them, visiting them and planning with the staff of the factory.

2. Planning in the classroom
Introducing the theme, including the subject areas and the methods involved; dividing pupils into groups to organise themselves and to select topics; an introductory talk on the factory, given by a member of the factory staff.

The groups prepare themselves for the visit.

3. Visit to the factory
Guided tour round the factory.
The groups start work on their particular topics.

4. Preparing the report
Each group prepares its own report.

5. Reporting
Reports, discussion and conclusions.

6. Evaluation
Evaluation together with the teacher, the pupils and the staff of the factory.
This forms the basis for the next visit.

Planning the next visit

The visits were carried out on one afternoon during the conference and were followed the next day by a workshop session at which case studies were developed. Each group was asked to develop a short case study to include the geographical location of the industry, a brief history of the factory and the processes within the industry. Other aspects of the industry were also considered, where appropriate, such as the economics of the factory, seasonal variation problems, wealth production, and social responsibilities of the industry such as training, labour intensiveness of the industry and any problems which had to be faced.

Visit to Chumundi Granites on the Bommasandra Industrial Estate

This granite cutting factory was built recently using modern machinery imported from West Germany with an 8.8 million rupee (about US$ 880,000) soft loan from the government; imported machinery was needed to meet the tolerances required by the export market. The firm employs 14 workers (25 staff in all) on an eight hour shift. It sells 6,000 square metres of granite each year at a price of US$120 per square metre. They used to ship the blocks, but now they process them first.

Granite from the quarry is cut into several cubic metre blocks with a 2 metre diameter saw, containing 128 segments tipped with industrial diamonds. The saw is loaded using an overhead crane. Slabs — 75 mm thick for (exported) monuments or 20 mm thick for wall-panelling and flooring — are cut according to the grain direction (depending on the visual effect required) and any large visible flaws in the blocks. This saw has a power of 60 kW. After the first cut, the slabs are ground (using 5 successive grades of grinder) and polished, and then cut to the required size with a 25 kW saw. Altogether the factory uses 175 kW of electrical power.

In the follow-up to this visit, participants noted the following in regard to planning a visit by students to a firm such as this:

(a) there should be prior correspondence with the industry, followed by a pilot visit by one or more teachers;
(b) during this preliminary visit, teachers should identify scientific principles and safety precautions taken, which could be highlighted during the visit by the students;
(c) handouts should be produced for the students (see below) together with instructions. The instructions should encourage them to consider technological, management and economic aspects, together with any by-products of the industry (in this case, "waste" stone used for mosaics).

It was agreed that such a visit should form a component of pre-service or in-service training for other teachers.

It was suggested that the handout to guide the students might include the following:

(i)   Chart the flow of materials in and out of the factory by drawing a flow-chart.
(ii)   Identify those sections of the flow-chart where humans interact with the flow.
(iii) Determine whether each stage of this system is labour or machine intensive.
(iv)   Take one section of the total flow and relate the activity to concepts which have been learnt in the classroom.
(v)   List three ways in which the management protects workers from hazard.
(vi)   Identify any potentially harmful by-products of the process.
(vii) In three sentences, describe the industry's contribution to the community in which it is located.

Visit to Ashok Fruits PVT Ltd

This food-processing firm lies 8 km east of Bangalore and was started 2 years ago with two major shareholders, Ashok Gulati and Totapuri Neelam. All the fruits processed — mango, pineapple, and papaya — are exported to the USA, the UK, the Gulf States and the USSR. The products contain no added colour or preservatives (apart from citric acid to enable a lower temperature sterilisation to be used) and they have a shelf-life of 2–3 years. The factory produces 20 to 40 tons per day on a 5 day cycle and has a storage capacity of 500 tons net. The cans are imported from Taiwan. The purchase price of the raw materials is from 1 to 2 Rs per kg.

The process involves collecting and ripening the fruit (in which protection from excess pressure is a consideration); cutting and slicing (with consideration of time, temperature, skills and enzyme action); mashing and pulping; cooking (where relevant); canning (automated) and sterilising the sealed cans; despatch; quality control through random sampling.

It was agreed that the visit would be suitable for the topic of food preservation in biology courses from the top of primary schools to adult education. Its purposes could be (i) to illustrate the scientific principles of preservation, (ii) to show students wider social and economic aspects of industry.

In preparation the students should be given a flow diagram for the process: perhaps they could be asked to sequence separate parts of it. Alternatively the students could be given the start and finish of the process and asked to fill in what goes on in the middle. A preparatory video would be helpful. Students going on a visit should be encouraged to ask questions, for example such leading questions as "how is the quality of the final product checked?". For follow-up, a classroom display is recommend with a discussion on the extent to which the industry meets social and economic needs.

Discussion on industrial visits

The visits to these factories in the Bangalore area were instructive. They were made particularly instructive by the quality and quantity of planning and preparation. The following points emerged from the discussion.

1. "Science and technology for industry" is one aspect of "science and technology for the community" since industry is part of the community and helps to ensure its continuing stability. It includes a concern for people, raw materials and the final product, and involves solutions to human problems. Students should be made aware of this and of the potentially symbiotic relationship between industry and the classroom.
2. An industrial visit may illustrate topics already taught in class or serve as a starting point for introducing new topics.
3. The Heinola model may be used as one way of describing pupils' visits to the industry. The model may be extended to include classroom practical work or other student activities at an appropriate stage. One example would be the use of a simulation-game *before* the visit.
4. A visit which is a follow-up to previously taught material may still be used to highlight new areas of possible classwork.
5. Involvement of schools with industries may be encouraged by schemes in which individual schools are linked to individual firms, as has been done in the UK.
6. There is a potential source of misunderstanding if students' experience one aspect of industry only. However, industrial visits may concentrate on a particular theme. For example, "measurement" has been used as a topic in one factory in which much use was made of vernier scales and other measuring devices.

Finally, it was agreed that it can be misleading to generalise practice between countries as between industries. The problems and potential solutions will vary between countries as also between industries.

Reference

1. Kurtenen, H. "Visit to Industry, based on the pupils' own activity and its diffusion in the Senior Secondary School in Finland". Symposium paper, volume 1. 3rd Int. Symposium on World Trends in Science and Technology Education, 1984.

34
Taking the Classroom to Industry

C. CHAMBERS

Bolton School, UK

When I first began teaching I found that most of the industrial visits I arranged were regarded by students with a mixture of excited anticipation and the joyful prospect of time out of the classroom. Industry, for the most part, was regarded by students with trepidation. Despite a great deal of effort and good will on both sides, the visits were often an anticlimax with disappointment and disillusionment on one side or the other.

Failures could be attributed to a number of factors, but primarily to a lack of knowledge of each other which led industrialists to pitch their talks at too high or too low a level of understanding for the students. Inadequate explanation of the processes being viewed, or mere inaudibility of the speaker were also factors in some cases. This period of contact building, however, produced a number of spectacular successes which ultimately led me to devise the specialised industry–education link scheme which I called the "Programmed Visit".

Programmed visits

Fundamentally a *programmed visit* is a lesson taught on industrial or commercial premises. The content of the lesson must be an integral part of the school curriculum with the subject matter chosen from an examination syllabus, whether devised internally by the school, or as part of a nationally accepted syllabus for an external examination. The objectives of these visits are:

(i) To indicate the relevance of subject material taught in schools to commerce, technology and industry;
(ii) to enable students to obtain a clear picture of industrial life with its opportunities and challenges.

The content of a *programme* will vary depending on the type of industry involved but will in essence be:

(a) an outline of the essential preparation work to be done in school by the teacher working with the pupils, including examination syllabus references where possible;
(b) a summary of the introductory talk given by a company representative to the students at the start of the visit;
(c) an outline of the site visit including each element of the tour;
(d) work sheets for students to use during their visit;
(e) post-visit material, including problems for homework, or participation in the subsequent classroom discussion of the visit.

In short, a *programmed visit* aims to enable a teacher to begin the study of a subject in the classroom, continue and reinforce this subject by a visit, and complete the work by further classroom discussion.

If the objectives of the programmed visit are to be achieved, a good deal of forward planning is essential, with work done by people on both sides of the link. This work can be divided into several distinct stages.

(i) The initial contact

Very early the backing of senior management is required, and it is to them that the initial approach must be made. If agreement in principle for a visit is reached, a teacher, or preferably, two, must visit the sites for a discussion of the nature of the company's activities. A tour of the site is valuable at this stage.

(ii) The subject of the visit

Frequently, several possible subjects emerge during the initial visit, and it is better to allow a little time for reflection before finally deciding which topic, or topics, are to be the subject of the visit. At this stage of the negotiations misconceptions on both sides of the link can be overcome, and a frank exchange of ideas pursued.

(iii) The initial programme outline

After the first meeting it is possible for the outline of the proposed visit to be prepared. This can be produced by either the teacher or the industrialist, but it is usually done by the teacher.

(iv) The trial visit

Once an outline programme has been produced and agreed by discussion between the teacher and the industrialist, a trial visit can take place. For this visit the teacher will frequently use students from his own school. Any problems which arise, or suggested improvements, can be considered after this visit and necessary modifications made.

(iv) The final programme

A programme draft is produced after the trial visit with all the necessary modifications and this, once agreed by both sides, becomes the final programme for the visit.

The scheme in operation

Once produced the programmed visit material is reproduced and circulated to all schools in the area. The name of an industrial contact able to arrange dates for a visit is included. Each company decides for itself the number it is prepared to accept, the dates when these visits can take place, and the number and ages of children permitted on a given visit.

Every school receives copies of programmes for industries in their area and is able to book directly. This enables students in schools to gain experience of a wide variety of industries. Arrangements can also be made for schools from further afield to visit these industries. The programmes available cover topics as diverse as corrosion and water treatment to company finance.

Advantages of the programmed visit scheme

These can be divided into two sections.

Advantages for industry

It is not the main role of industrialists to educate, and if they do agree to give up some of their valuable time for this work it must be time well spent. Several important points can be made in favour of the programmed visit scheme:

(i) visits are meaningful, aimed at the appropriate level of understanding, and students are well prepared in advance;

(ii) individual planning is not necessary for every visit;

(iii) since visit material is well documented, the same staff do not need to be used for every visit;

(iv) staff involvement in planning visits is limited to the period of the programme production;

(v) students gain a valuable experience of the industry visited.

Advantages for schools

(i)   Industry is well prepared for the visit;
(ii) students will be able to understand the content of the visit;
(iii) preparative work for the visit is well defined and limited to syllabus material which is relevant to student courses;
(iv) worksheets ensure student involvement in the visit;
(v) visit can be arranged on a range of different subjects;
(vi) the work of a few dedicated teachers and industrialists can benefit a lot of students in many schools.

Essential features of a successful scheme

Just like any successful business or school activity, this education–industry link scheme needs enthusiasts with energy, initiative and enterprise, who are willing and able to devote time to help others. This scheme was devised by two teachers with the support of an influential industrialist, who was able to convince senior management in other companies that the proposed scheme was sound and worthy of their consideration. This single action enabled the organisers of the scheme to gain access to a wide range of companies.

A conference was held, supported by the Royal Society of Chemistry, the Department of Trade and Industry and a number of companies to involve more teachers in programme production. The scheme now has wide support and is working well in many areas.

The rapid technological and educational changes which are occurring are making it imperative for teachers to prepare students for a future which will require them to be both flexible and adaptable. The *programmed visit* industry–education link scheme enables students to gain wide experience of industry in a relatively short time. Students may not like what they see, but certainly they will be better able to appreciate how other people earn their living. If this brings tolerance and understanding to our fellow men, then indeed, our efforts will not have been in vain.

35

An Alternative to Industrial Visits: a Project in Thailand

LADDAWAN KANHASUWAN

Pranakorn Teachers' College, Bangkok, Thailand

Teaching materials for the study of Thai industry for inclusion in the curriculum were developed in Pranakorn Teachers' College to help secondary schools to overcome shortcomings in this direction.

When teaching a particular concept, the teacher must be careful in selecting the appropriate approach, as well as teaching aids and media. Some concepts call for demonstration, others require experimentation. However in teaching about industry, it is often not possible for the teacher to demonstrate an actual industry site nor for the pupils to carry out an experiment. The best method is to take pupils to a factory and to hear lectures by the factory staff, as discussed elsewhere in this book. Unfortunately, a large number of teachers are unable to do this. Schools are unable to provide transport and pupils cannot afford fares; the number of pupils is too great for factories to accommodate; some schools are too far from the factories. Thus an alternative strategy must be adopted.

The Industry Project has therefore developed Industrial Process Model Kits, together with slide-tape programmes, for classroom use in place of actual visits to industrial sites.

Producing the kits

The kits are produced as part of a course for third-year science students at a teachers' college, who are divided into groups of six. Each group decides on the kit to be produced, the choice of subjects coming from discussions with teachers, and each group then spends about two months studying all aspects of the factory, analysing the stages of production. Slides are made and samples of raw materials and products of each step are collected.

The Industrial Process Model Kit is constructed so that it fits in a box for easy transport. Information sheets and evaluation tests are written for the teacher to use and, if the necessary equipment is available, a slide-tape programme can be produced. Through each stage of the work by the student group, staff members from the factory give advice. When a kit is completed, it is classroom tested in secondary schools. Then any weak points are improved and perfected before it is produced in final form.

Since the beginning of the project in 1980, twenty-three Model Kits have been developed, of which the following are examples:

1. Thai silk
2. Thai bricks
3. Sugar from coconut flowers
4. Bronze ware
5. Paper from rice stalks
6. Glassware
7. Water supply
8. Solid fuel
9. Mae Khong distillery
10. Singha Beer brewery

To publicise the kits, which can be lent to schools, students take them to schools where they are on teaching practice. They are also shown at exhibitions.

Using the kits

One successful technique for using the kits in a school is to divide the class into groups of five. The six parts of a kit are laid out in different parts of the room, with a seventh station for a slide-tape presentation. Each group of pupils is assigned to a station and, when finished, rotates to the next station until each group has studied all six stations. Any group which finishes early and has to wait for the change of station can use the slide-tape at the seventh station.

Industrial and Technological Issues in the Secondary Science Curriculum: Assessment

Introduction

Assessment of economic, industrial, technological and social issues in science calls for skills, for the teacher writing the questions, for the student answering the questions and for the teacher correcting the answers, which are different to those usually associated with writing, answering and correcting questions on scientific projects.

Thomas addresses these problems in the context of the science course "Physical Science — Man and the Physical World", taught in the state of Victoria in Australia. He examines, in particular, the nature of the objective being assessed, the constraints of a public examination system, and the need to make items relevant in context and yet reflect the content of the syllabus.

One of the most difficult things to do is to write questions that both test the underlying scientific principles and reflect the aims of the course. For it must be realistically accepted that change in curricula will not occur without change in examinations, and that examinations will need to reflect the industrial context in the same way that industry is reflected in the curricula themselves. Such a requirement implies a substantial move from the traditional methods of assessment currently encountered.

Finally, in this section, examples of assessment questions from several of the courses discussed in this book are given which will, in themselves, give further ideas and raise issues within the minds of our readers.

36

Assessing Understanding of Science, Technology and Society Interactions in a Public Examination

I. D. THOMAS
Monash University, Australia

In 1974, a group of science educators in Victoria, Australia began developing a new science course (Physical Science: Man and the Physical World) for the Higher School Certificate, a course which would be assessed by public examination. It is based around three core topics: Change and Interaction; Useful Materials; and Energy Transformations. To quote the Victoria Institute for Secondary Education, each of these topics is

". . taken from the point of view of science and technology, their effects on people and the effects of people on the development of science and technology."

The achievement of these goals depends to a large extent on the way in which the subject is taught and assessed, and the examination exerts a considerable influence on classroom practice. To give some feeling for the way the course attempts to achieve the desired balance between science, technology and society, I have included some information about a topic in each of the core units, together with sample examination questions.

Example 1: Change of momentum

The general thrust of the Change and Interaction Unit is concerned with the identification and degree of control which one can exercise over variables. The situations investigated begin with straight line kinematics in which identification of variables is straightforward and considerable control can be exercised over variables. A more complicated situation is investigated in relation to chemical changes. Finally there is the investigation of change in society, where control of variables is very difficult.

The topic, change of momentum, allows consideration of simple calculations involving an understanding of the relationships

$$p = mv, \ \triangle p = m\triangle v \text{ and } F\triangle t = m\triangle v$$

leading into investigations of Newton's second law of motion and the relationship

$$F = ma.$$

In practical work, students investigate some of these relationships through observation and measurement with trolleys, including investigation of simple collisions in a straight line which involves an understanding of conservation of momentum.

Technological aspects can be covered in topics such as space flight, including launch re-entry phases, as well as movement of the vehicle in outer space; railway shunting; motor vehicle collisions, etc.

Space flight topics naturally lead to questions of value, involving the huge expenditure of resources and the potential for waging war from outer space. Discussion of motor vehicle collisions opens questions of road safety, factors involved in the road toll and the type of research that is done as a result of public concern about road safety. The following are two typical examination questions.

Question 1

The motor car is a complex of technology which has an enormous impact on society. The motor industry is a key element in the Australian economy because of the number of people it employs directly or indirectly.

An important negative impact of the motor car is the incidence of road accidents in which over 3000 Australians are killed and many thousands more injured or maimed **every year**. Research has shown that speed and driving under the influence of alcohol are key causes of road accidents. These two factors are brought to focus in the **reaction time** of a driver faced with an unexpected hazard (i.e. the time between seeing a hazard and the initiation of any avoiding action, e.g. putting a foot on the brake).

a.   A fit, young, sober driver has a reaction time of 0.75 second. A motor car travelling at 60 km hr^{-1} will travel a distance of 12.5 m in this time.

A tired, intoxicated driver has a reaction of 1.5 second. How far will his car travel during this reaction time if he is driving at a speed of 120 km hr^{-1}?

b.   Suggest TWO measures which might be taken to reduce the incidence of motor vehicle accidents through speed and intoxication and thus reinforce the responsible use of motor car technology.

Question 2

In a programme to test safety features of its cars, a manufacturer arranges controlled "crashes" of a sample of vehicles from the assembly line. In the tests a car of mass 1500 kg is crashed into a solid wall (i.e. a wall which does not move or suffer any damage in the collision) at a speed of 20 m s^{-1}. The collision lasts 0.2 second; i.e. the time from the first contact with the wall until the car comes to rest is 0.2 second.

a.   What is the momentum of the car before the collision?

b.   i.   What is the force exerted by the car on the wall?

    ii.   What is the force exerted on the car by the wall?

c.   In another test, a similar vehicle (of mass 1500 kg) but constructed of panels which crumple progressively (like a concertina) on impact, and with a specially designed shock absorbing chassis, is crashed into a solid wall at a speed of 20 m s^{-1} and takes 0.3 second to come to rest.

Other tests involved vehicles containing dummies with and without seat belts. In controlled crashes the dummies without seat belts were thrown about inside the car while the dummies wearing seat belts remained in their seats and took 0.5 second to come to rest.

    i.   What key variable in a collision is being manipulated in these tests involving seat belts and the progressively crumpling panels of this vehicle?

    ii.   What do you consider a disadvantage of having panels which crumple progressively on impact in a collision?

d.   The "solid" wall that helps control the smashes is an important feature of these tests. In the crashes there is an apparent loss of momentum from the car-wall system. Briefly explain how this apparent loss of momentum occurs.

Question 3

In launching a space shuttle vehicle like *Challenger*, both solid and liquid fuel rockets are used. The liquid fuel component, consisting of hydrogen and oxygen, burns for 8 minutes and produces an average thrust of 5×10^6 N.

a.   What is the impulse given to the *Challenger* launch vehicle by the combustion of the liquid fuel?

b. Give a brief explanation for the small initial acceleration of the *Challenger* launch vehicle, in spite of the enormous thrust generated by the rockets.

....

....

c. It has been argued that the space programmes of the major powers are an enormous waste of money and resources, and that the funds could have been better directed towards humanitarian projects. Consider this proposition, then:

i. indicate ONE benefit that has flowed from the space programme.

....

....

ii. indicate ONE problem that faces mankind as a result of the space programmes.

....

....

iii. indicate whether you agree with the proposition or not and then add an additional point to support your position.

....

....

Example 2: Efficiency

The unit on Energy Transformations attempts to stress the fact that in transformations the quantity of energy is conserved, while the quality is not. Some important energy conversions are examined in terms of efficiency.

The treatment of efficiency begins with definition and simple calculations with examples covering various machines and energy conversions. Practical work involves simple observations of a small motor lifting a load.

Discussion can centre around ways of improving efficiency, for example, in fireplaces, refrigerators, air conditioners and engines. This leads to the notion of saving (or not wasting) energy and the costs of inefficiency. This discussion can lead into consideration of alternative energy sources (for example, solar energy) for particular applications. An important area of discussion is energy around the home.

Question 4

The recently completed Newport Power Station in Melbourne has a power output of 500 MW of electricity. This is a natural-gas or oil fired station sited on the banks of the Yarra River about 5 km from the centre of the city.

a.   If the overall efficiency of the station is 40%, what is the rate of energy input (gas or oil) to the station burners which heat the steam which turns the turbine to produce electricity?

b.   If burning natural gas produces $40\ MJ\ m^{-3}$ how much gas is used each minute when the power station is operating at full capacity?

c.   State TWO positive reasons for establishing such a major power station in the Melbourne area.

d.   State TWO criticisms that opponents of the Newport Power Station might make of the station now that it is in operation.

Example 3: Steel and wood as building materials

In this unit on Useful Materials, attempts are made to relate properties to the internal structure of the material and also to the possible uses of it. Part of the discussion can then go on to consider other alternative materials for similar functions, again from the point of view of the internal structure of the material, for example the use of aluminium or concrete for building frames.

Discussion would have to consider the criteria for selecting a material for a particular job, which must include consumer preference and aesthetics, as well as utility or other special criteria (for example, fire resistance properties, an important issue being discussed in the community following the disastrous bush fires in S.E. Australia in 1983).

Question 5

Write a short essay outlining the properties and uses of wood as an engineering material. You should

● describe the structure of the material (a labelled diagram would help);

● comment on its qualities that make it a suitable material;

● comment on its limitations.

You may wish to compare wood with steel in their uses in building bridges and house frames.

Problems in assessing student understanding of STS interactions

Let us now turn our attention to examining the STS component of the subject. Most of the problems encountered can in some way be traced back to the fact that in the assessment of student understanding of STS interactions we are dealing with very complex behaviour. Some of the behaviours to be measured include knowledge of and acceptance/rejection of attitudes; values; approaches to problem solving, knowledge generation and data handling; the development of personal value systems and attitudes; the analysis of situations to ascertain what competing value systems are operating and the evaluation and interpretation of data used in supporting opinions and arguments.

The measurement problems inherent in assessing behaviour of this complex nature are well known, but bear repetition. We cannot operate from the position that there is any single correct opinion, attitude or value. There may be commonly accepted opinions, attitudes or values but this neither makes them "correct" nor superior or to be preferred to alternative opinions, attitudes or values. Thus there is a great deal of subjectivity introduced into any assessment and there is considerable difficulty in ensuring comparability of responses. Various techniques like panelling, negotiation and multiple marking of responses go some way to removing this problem. Nevertheless examiners need to remain open to the possibility of an unexpected (yet relevant) point of view being expressed and argued within an unusual set of values and attitudes.

Obviously students will be at different stages in the development of their personal value systems and will also be operating from different experience and knowledge bases and we have no way in the one shot public examination of gauging how much development (if any) has occurred during the year. Further, there is no sure way of assessing whether any changes or growth can be attributed to the course itself or to some other influence. In the external examination we can only obtain a still-frame view of a very dynamic situation, knowing that there are a host of factors that could operate to cause distortion of that fleeting image.

In recognition of these measurement problems the Physical Science subject committee has successfully argued that the more appropriate means of measuring some of these behaviours is to vest the responsibility in the classroom teacher who has a much more intimate contact with the students over eight or nine months. This responsibility is backed up by a moderation scheme which both checks on the assessments given by the classroom teacher and provides a means of sharing and learning from peers facing the same problems. Fifty per cent of the end of year assessment is based on the school assessment of a student. Our experience shows that the correlation between school assessments and the external public examination are sufficiently small to indicate that a different set of behaviours is being assessed in the two approaches. This is not to say that we have necessarily been successful in devising approaches to measure all of the behaviours that are relevant to this subject.

Some of the objectives concerning STS interactions require the use of test items that go far beyond simple knowledge recall. In effect the test items must test for lateral transfer of knowledge and skills to a parallel situation. The problem is that once a particular situation has been used it becomes part of public knowledge and thus its subsequent use becomes essentially a knowledge recall item.

Since we are testing for transfer this in itself imposes a requirement that teachers employ teaching strategies that facilitate the transfer of learning — i.e. that emphasise the generalisations rather than the particular. This is an important and distinguishing feature of Physical Science in comparison to the more traditional teaching of other Science subjects.

In terms of setting or writing test items it therefore becomes necessary to provide quite a lot of information (even to do a little teaching) in a question stem so that students can perceive the parallels between the test situation and the specific learning experiences they had in the classroom. This can pose considerable problems in the area of technology where there is often a specialised set of vocabulary, jargon and procedures that have developed with the technology and are largely incomprehensible to the outsider. Question 6, relating to the wine industry, is a good example of this.

Question 6

Depending on the conditions established by the winemaker, grape sugar ($C_6H_{12}O_6$) can produce either wine or vinegar. The chemical equations describing what happens during the chemical reactions are as follows:

Production of wine

$$C_6H_{12}O_6 \xrightarrow{\text{yeast}} 2\ C_2H_5OH + 2\ CO_2\ (g)$$

$$\left(\begin{array}{c}\text{aq. solution density} \\ 1.10\ \text{g cm}^{-3}\end{array}\right) \qquad \left(\begin{array}{c}\text{aq. solution density} \\ 0.98\ \text{g cm}^{-3}\end{array}\right)$$

Production of vinegar

$$C_6H_{12}O_6\ (\text{aq. solution}) + 2\ O_2\ (g) \xrightarrow{\text{yeast}}$$
$$2\ CH_3COOH\ (\text{aq. solution}) + 2\ CO_2\ (g) + 2H_2O\ (l)$$

a. What does the winemaker do to prevent his wine from becoming too vinegary (acid)?

....

....

b. How could the winemaker monitor the production of wine to check on the amount of acid (CH_3COOH) in his wine (i.e. what test could he perform)?

....

....

c. Assuming he has established the correct conditions for wine production, how could the winemaker monitor the progress of the wine making process?

....

....

d. What function is the yeast fulfilling in the chemical reaction that leads to the production of wine?

....

The dual requirements of testing for transfer and the need to provide students with an adequate and comprehensible data base often results in a considerable reading load which is incompatible with the time restraints of an external examination. There is a real chance of oversimplifying a situation so much that the reality of the exercise is diminished and the level of response is trivialised.

Question 7

On December 19, 1982, "The Quiet Achiever" left Perth to travel about 4000 km across Australia to Sydney, powered by solar energy. The journey was completed on January 7, 1983 after 172 hours of running time at an average speed of 7 m s^{-1}.

"The Quiet Achiever" is constructed of a light tubular space-frame supported by four racing-bicycle wheels. It is encased in a streamlined fibre glass cocoon and topped by a flat table top of 20 solar (photovoltaic) panels, covering an area of 8.5 m^2, and providing up to 600 watts of power in direct sunlight.

The solar cells were wired to supply 24 volts to two standard automobile storage batteries connected in series. The 24 volt, 650 watt electric motor was powered from the batteries and provided a maximum speed of about 60 km hr^{-1}.

Power was transmitted to the wheels via a four speed chain transmission, while the brakes were standard bicycle brakes. The total weight of the vehicle (including the driver) was 200 kg.

a. Indicate the sequence of energy transformations that enable "The Quiet Achiever" to travel along the road in sunlight hours. (A diagram may be an acceptable answer to this question.)

b. How much kinetic energy is possessed by "The Quiet Achiever" when travelling at its average speed?

c. What electric current is delivered by the solar cells to charge the storage batteries, when the solar cells are operating at maximum output?

d. Will the application of technology demonstrated by "The Quiet Achiever" become commonplace in society in the next 20 years? State TWO reasons to support your opinion.

Question 8

Product bar codes like the one shown below now appear on many items stocked in supermarkets.

The first two digits (93) identify the country of origin as Australia; the next five digits (12345) identify the manufacturer; the next five digits (67890) uniquely identify the product by:

● type, flavour, etc.

● size of retail pack

● special or normal sale

● model or modification etc.

The last digit (7) provides an internal check of the information contained in the bar code.

The checkout assistant uses a scanner (most likely a laser beam code reader) which feeds electronic signals to a microcomputer to

● read the bar code

● retrieve the item description from memory

● send the price and description to the electronic cash register which provides a visual readout for the customer and prints the itemised docket tape.

The price of the item is entered by the retailer at the local store. The local store computer also maintains a continuous stocktake of items on the shelves and in the store and automatically re-orders products. The marketer of the product bar code system claims that its use is of benefit to the consumer, retailer, store employee and manufacturer. Others suggest there are both real and potential disadvantages to retail store employees and customers.

Write a brief argument in favour of (or against) the general introduction of product bar codes in retail stores.

In writing your answer you might consider:

● advantages and disadvantages to the customers

● advantages and disadvantages to the retail store employees

● advantages (or disadvantages) to the retailer

● ways in which the retailer can overcome the disadvantages to customers and store employees.

Relevance

The objectives for Physical Science require that what students learn in school has relevance to the real world. This requires that the examination should attempt to deal with current real-life situations as well as the idealised systems of academic science (classical physics and theoretical chemistry). There is a conflict in this situation also with the realities and constraints of the external examination system. The preparation of an external examination paper involves lead times of from 6 to 18 months and this can result in many exciting, relevant and controversial issues being left out of examination papers. Teachers in the schools do not suffer the same constraints and they should be alert to the many opportunities provided in various newspaper, magazine and television reports which give a different and contemporary slant on the physical world (albeit not necessarily scientifically accurate). Question 8 is a recent examination question used in Physical Science. The example on the bar code pricing technology was a case in which the examiners had published the item before it became a popular issue in the media.

Constraints imposed by the public examination

Some of the constraints of a public examination have already been raised in this paper. The most obvious constraint is that of time i.e. the length of the examination. Time also features in the amount of reading (thinking) necessary to form a response and then to write it. Other constraints include the requirement to maintain a balance between various skill and content objectives or of ensuring an adequate and properly weighted coverage of the syllabus; use of test items with an appropriate level of difficulty so that an adequate spread of marks is obtained to enable the examination to serve its primary purpose as a rank ordering screening instrument; the use of test situations which are amenable to pencil-paper responses; and giving candidates a fair chance to demonstrate what they know and have learned.

It is clear from what has been written here that some of the objectives relating to STS interactions can be best achieved through assessment in the classroom and not by external examination. The procedures established for Physical Science provide the climate and incentive for teachers to attempt the type of measurement necessary for this purpose. Some attempt has been made to include test items on this area in the external public examination. Sample test items have been included in this paper. More examples can be obtained by purchasing copies of past examination papers from the Science Teachers Association of Victoria. Sample answers to these examination papers are available from the same source and these give some indication of the level of reasoning examiners are prepared to accept from candidates.

A challenge

Assessment in the area poses an interesting challenge to the professional teacher. The techniques required are similar to those needed for assessment of attitudes and problem solving. The school situation is clearly well suited to measurement and assessment of objectives related to STS interactions and the challenge is for teachers to take up this professional responsibility. The challenge for rigidly controlled public examination systems is to become free enough to enable measurement of some objectives to be made at the place most appropriate for that measurement i.e. the school classroom; and to build in procedures that both protect the classroom teacher and encourage him/her in performing this important task.

37
Further Examples of Assessment Items

The Science and Technology in Society project

This UK project provides resource materials which relate industrial, technological, social and economic aspects to the science topics in the normal school curriculum. It aims at pupils aged 13–16. In addition to the resource units being produced by the project, work is also being done on the development of assessment items relating science topics to real situations. Some examples are given below. Note that the questions are assessing conventional science knowledge and skills, but they are set in an everyday context and also expect some awareness of the economic and social implications of science.

One problem with setting questions in an industrial or everyday context is the need to provide the student with the necessary information relating to that context: it would be unfair to expect students to be able to recall detailed information relating to a wide range of contexts, and in any case this would limit the contexts which could be used to those listed in the syllabus. It is therefore important to provide the necessary information in the question itself, condensed as much as possible.

Question 1 (based on "Chemicals from Salt")

Sodium hydroxide ($NaOH$), chlorine (Cl_2) and hydrogen (H_2) can all be manufactured by the electrolysis of sodium chloride ($NaCl$).

a.   In what state must the sodium chloride be when it is electrolysed?

b.   At which electrode is chlorine produced in electrolysis?

c.   Some of the products of this process can be used to manufacture hydrochloric acid, HCl (aq).

    i.   Which two products of the electrolysis of salt can be used to manufacture hydrochloric acid?

    ii.   Outline how you would convert these products to hydrochloric acid.

iii. If a company makes hydrochloric acid in this way, it will have to find a market for *two* chemicals. Hydrochloric acid is one. What is the other?

A company called BRINETECH is involved in the production of chemicals from salt. Brinetech sells most of its products to the following buyers.

Sodium hydroxide – to a soap manufacturer.
Chlorine   – to a PVC manufacturer.
Hydrogen   – to a margarine manufacturer.

d. If the PVC manufacturer goes out of business, Brinetech must find a new market for its chlorine at once. Explain why, and suggest one possible industrial use for Brinetech's chlorine.

e. If the margarine manufacturer goes out of business, Brinetech may be able to use the hydrogen in its own factory. Suggest one way the hydrogen could be used.

f. Brinetech makes 200 tonnes of chlorine per day. Calculate the mass of sodium hydroxide they make per day.

Question 2 (based on "Electric Vehicles")

The following short passage is about electric vehicles. Read the passage, then answer the questions that follow it.

There are two major problems preventing electric vehicles making a major contribution to road transport. One is mass, and the other is recharging time. Electric motors are 5 times heavier than equivalent petrol engines, and batteries further increase the mass of an electric vehicle. Electric vehicles can only cover 50 miles or so before their batteries need recharging, and recharging takes several hours. These two factors give electric vehicles a severe disadvantage compared with petrol or diesel vehicles.

a. Suppose an electric vehicle accelerates from rest to a speed of 45 km/h. It runs steadily at this speed for 2 minutes, then its brakes are applied and it comes to rest.

i. Explain why the vehicle must use energy to accelerate.

ii. Explain why it needs energy to run at a steady speed.

By the time the vehicle comes to rest, it has used up some of ̣he electrical energy stored in its batteries. What has this ̣gy been converted to?

b. The extra mass of electric vehicles means they must consume more energy than petrol vehicles for a given trip. Explain why.

c. A battery charger supplies electricity at 12 volts. Its maximum power is 2 kW. What current does it deliver at maximum power?

d. An electric vehicle's batteries store 40 000 kJ of electrical energy. How long would a 2 kW charger take to recharge them from fully discharged to fully charged?

e. In spite of the disadvantages mentioned earlier, electric vehicles have several advantages over petrol and diesel vehicles. Suggest *three* advantages you think an electric vehicle would have.

Question 3 (based on "Controlling Rust")

The following list gives a number of ways that can be used to prevent the rusting of iron and steel.

    A   painting
    B   covering with oil
    C   plating with zinc
    D   plating with chromium

a. Explain why covering or coating an iron surface prevents it rusting.

b. Which of the methods would give the *most* effective protection against rust? Explain your answer.

c. Which method would give the *least* effective protection against rust? Explain your answer.

d. Zinc coating continues to protect iron even when part of the coating has been scratched. Once a coat of paint has been scratched, on the other hand, it no longer gives protection. Explain the reason for the difference.

The costs of different methods of protection are as follows.

    A   painting — medium cost
    B   covering with oil — low cost
    C   plating with zinc — medium cost
    D   plating with chromium — high cost

For each of the following jobs, say which protection method would be most suitable. Give a reason for your answer in each case.

e. Protecting a corrugated iron roof on a shed.

f. Protecting steel sheeting that is to be pressed to make car bodies.

g. Protecting a steel filing cabinet.

h. Protecting iron fencing wire.

Question 4 (based on "Recycling Aluminium")

This table gives some information comparing three metals — aluminium, gold and iron.

	aluminium	gold	iron
Percentage abundance in the Earth's crust	8.1%	0.0000004%	5.0%
Approximate price/ £ per tonne (1984)	900	10 000 000	130
Annual world production/tonnes	14 000 000	1 400	450 000 000
Time known reserves will last/years	30	27	302
Position in the reactivity series of metals	above iron and gold	below aluminium and iron	below aluminium but above gold
Malleability (how easily the metal can be hammered into a thin foil)	good	very good	fair
Percentage of the metal used which gets recycled	35%	over 90%	50%

(a) Most metals, including aluminium and iron, do not occur on their own in the Earth's crust. They occur as ores. What is meant by an "ore"? (1)

(b) Gold is one of the few metals that occurs on its own, uncombined, in the Earth's crust. Why? (1)

(c) Aluminium is more abundant in the Earth's crust than iron. Yet aluminium metal is much more expensive than iron. Explain. (2)

(d) Aluminium is more reactive than iron. Yet iron corrodes (rusts) easily, while aluminium does not. Explain why. (1)

(e) Aluminium is used for household foil, for wrapping up and covering food. Iron is never used in this way. Give *two* reasons for the difference. (2)

(f) Explain what is meant by "recycling" a metal. (2)

(g) Give *two* reasons why it is important to recycle a metal like aluminium. (2)

Give *two* difficulties which have to be overcome before we can ~ycle more aluminium. (2)

~o reasons why much more gold is recycled than the other ¹s. (2)

Question 5 (based on "Industrial Gases")

This question is about industrial gases — gases which are produced industrially in large quantities.

The table gives some information about five important gases.

Name	OXYGEN	NITROGEN	HYDROGEN	ARGON	PROPANE
Formula	O_2	N_2	H_2	Ar	C_3H_8
Relative molecular mass	32	28	2	40	44
Percentage abundance by volume of gas in air	21	78	0.001	0.9	very small
Does it burn?	no	no	yes	no	yes
Does it let things burn in it?	yes	no	no	no	no
Approximate cost of gas (£ per cubic metre)	0.50	0.50	1.50	2	0.90

Using the information in the table, and your own knowledge of the gases, answer these questions.

(a) Some of the gases are manufactured by the distillation of liquid air. Which ones?

(b) For the gases that are *not* manufactured from liquid air, give *one* major source of each gas.

(c) Which two gases could be used together to propel space rockets?

(d) Write an equation for the reaction that happens when the gases in (c) are used as a propellant.

(e) Which two gases could be used to manufacture ammonia?

(f) Write an equation for the manufacture of ammonia from these two gases.

(g) Which gas could be used to fill lighter-than-air balloons?

(h) What is the safety hazard of using this gas in balloons?

(j) Suggest one reason why argon is more expensive than nitrogen.

(k) For each gas, give one major use of the gas not already mentioned in this question.

N.B. This question could be made harder by omitting some of the information in the table — e.g. percentage abundance in air, information on burning. But the question would then rely on greater factual recall.

**Further examples from the Australian project
"Physical Science: Man and the Physical World"**

Question 1

Dr Paul Wild, Chairman of CSIRO, suggested in June 1984 (*The Australian*, 23/6/84) that a high speed electric "bullet" train could run between Melbourne and Sydney, east of the Snowy Mountains and by way of Canberra and Gippsland. The journey of 1000 km would take approximately three hours travelling at speeds up to 400 km hr^{-1}.

(a) i.   What do you think would be the best source of energy for the train?

....

....

ii.   Justify your answer.

....

....

(b) i.   Dr Wild claimed that the 100 tonne train would succeed in maintaining its speed even in the hills, since it would climb them by using its own kinetic energy and momentum. If the train approaches a hill at a speed of 360 km hr^{-1} (100 m s^{-1}) what will be its kinetic energy?

ii.   What will be its momentum?

(c) The train climbs a 100 metre high hill. Consider no forces other than that of gravity and assume the acceleration due to gravity is 10 m s^{-2}. How much kinetic energy will be converted to gravitational potential energy?

(d) Actually some energy is lost to heat through friction especially on bends, and the motors also increase the train's energy, so that it is found that the train's kinetic energy at the top of the hill is actually 405 MJ.

i.   What will be the velocity of the train at the top of the hill?

_____

_____

ii. Thus, what is the percentage reduction in speed during the hill climb?

_____

_____

(e) i.   What do you see as being some of the benefits of this advanced technology? Give TWO benefits.

....

....

....

ii. Do you think there would be any negative aspects which should be evaluated before proceeding with the project? Give ONE negative aspect.

....

....

....

Question 2

Two new methods of transmission, fibre optics and satellite relay, are being considered in Australia for the transmission of domestic television programmes, now mainly carried out by conventional means.

Fibre optic cables are thin glass tubes through which laser beams carry the programmes. Many programmes can be carried long distances.

In **satellite relay** (AUSSAT) the satellite remains stationary over one point on the earth's surface. The programmes are beamed up to the satellite on which "transponders" receive, amplify and then retransmit the signals back down to earth.

Conventional transmission relies on a combination of coaxial cables and very high frequency (VHF) or ultra high frequency (UHF) electromagnetic radiation from elevated transmitting antennas.

(a) Choose the most appropriate form of transmission for home viewers in a new urban area which has geographical characteristics like Melbourne.

....

Give one reason for your choice

....

....

(b) Choose the most appropriate form of transmission for home viewers in remote areas.

....

Give one reason for your choice.

....

....

An example from the New Zealand course "Science for all" for grade 9 students

(a) The three raw materials needed for the production of sulphuric acid are ...., ..., and ....

(b) Write a balanced equation for the overall reaction of the sulphuric acid plant in which the three raw materials form sulphuric acid.

(c) Two reasons for the low cost of sulphuric acid are .... and ....

(d) The burning of sulphur and the conversion of sulphur dioxide to sulphur trioxide involve the process of .... The conversion of sulphur dioxide to sulphur trioxide is carried out with the help of .... The temperature is important at this stage; if it is too high .... and it it is too low

....

Two reasons why efficient conversion is important are .... and ....

(e) List the stages in the process where heat energy is released.

(f) The thermal generating plant at Huntly uses 1000 tonnes of coal each day. If the coal contains 0.3% sulphur and all of this sulphur becomes sulphuric acid in the atmosphere, how much sulphuric acid is released into the atmosphere daily?

(g) In spite of the use of filters a sulphuric acid plant releases about the same amount of sulphur dioxide into the air as would be released by a hundred coal burning domestic fires.

What are the advantages and disadvantages of the sulphuric acid plant from the point of view of atmospheric pollution, compared with using the site for dwellings.

(h) Superphosphate is a mixture of .... and ....

(i) Why is rock phosphate not suitable as a fertiliser?

Education and the World of Work

38
The Needs of Industry

A workshop was held in Bangalore to explore the specific needs which industry has for the education and training of its future manpower and to place those needs in the context of general education.

The first speaker was S. G. Ramachandra, Chairman and Director of several engineering companies in Bangalore and of state committees concerned with engineering. He made the point that emphasis on higher education and technical training had provided many Third World countries with the required manpower with skills in engineering and management. However the bulk of the manpower needed by industry came from the formal secondary school system. The dilemma in India is twofold and it is no doubt the same in many developing countries. First, for every 100 children enrolled at primary level, over 90 drop out of the system before they reach the age of 17 and it is these drop-outs who are the recruits to industry. Secondly, the formal system is still based on the British educational system, which is inappropriate to developing countries at the end of the twentieth century.

The problem centres on the needs for manufacturing and operational skills which are not provided by the formal system. The contrast between the formal and non-formal systems of education is evident: among those apparently disadvantaged drop-outs are some who become entrepreneurs, who learn skills on the job and who set up service and manufacturing facilities. The non-formal system has reacted faster to changing technologies, both in urban and rural areas.

The vast spectrum of skills required in a Third World country like India — from bullock-cart operation to airbus technology, from a village postal runner-cum-messenger to satellite-based communication technology — shows that a highly centralised training scheme, uniformly applicable all over the country, is inappropriate. Nevertheless, industry feels the inadequacy of the educational system and seeks change. At present, industries have to develop their own training programmes which have no links at all with the formal education system. There is a wide gulf between industry and education, and there is a need in the Third World to reduce it.

The student of today will live in the environment of the technology of the next century. This fact must be appreciated by teachers and consequently efforts must be made to develop attitudes which will enable the student to expect change and to be responsive to it. It would help if school teachers had exposure to industry so that as many as possible in the primary and secondary system could gain experience of the industrial environment and its needs. Industry must contribute to this, extending its own programmes to educate teachers and to develop teaching aids and new methods in order to improve the quality of students leaving the schools who are seeking an opportunity for a career in industry.

The second contribution to the workshop came from R. D. N. Somerville and A. J. P. Sabberwal, respectively a divisional chairman and a head of research of Turner and Newall, a large international manufacturing organisation based in the UK, but with many companies operating in the Third World. They identified some of the needs which industry looks to science education to fill.

They stressed that the creation of national wealth to meet human needs depends on the success of the nation's industry. Industry operates in a scene of rapidly increasing technological advancement and international competition. This is true both in the advanced economies and in many of the developing countries, and successful industrial concerns must meet customer needs in an international context. An example of this is the rapid development of the textile industry in the Far East based on meeting the sophisticated tastes of the European and American markets. The identification of quality and design by the Japanese in cars and electronic products is a winning aspect of their rapid industrial growth. Less advanced developing countries have provided national wealth by encouraging small industries to meet world market demands.

Industry needs its share of scarce manpower resources. Unfortunately it has failed to project a favourable image to society. As industry works to improve that image, there will be a great need for a sympathetic response from education.

The following points summarise aspects of education of concern to British industry: it would seem that many of the lessons from the UK may have international significance.

1. Science and technology must be a *core subject* in the school curriculum both as a part of general education and a basis for subsequent vocational training. The curricula must prepare students for the realities of the industrial application of science.
2. Industry attracts a small share of the most able, but as Mr Ramachandra pointed out, industry's needs are met from men and women with a *wide range of abilities*, and this must not be overlooked. The person who is a so-called academic failure in the school system is often not a failure within industry. For this reason, assessment techniques at school must be devised which *encourage* students and which allow industrial management to select those who will thrive in an industrial context.

3. Industry must encourage *more women* into areas of engineering and other technologies.
4. Industry must improve the *status of engineers* and ensure that they command top management jobs as well as being very well rewarded when performing key specialist functions.
5. A much neglected aspect of engineering is the *maintenance* of technology at all levels. In developing countries this can be the most vital element after technology has been transferred. Cases abound of national economies being crippled by the breakdown of electricity generation plants, etc.
6. There is a special need for identification of the *respective roles* to be played by secondary education, tertiary education, technical training and industry in ensuring an adequate supply of engineering skills.
7. *Information technology* has to be seen as a new industrial revolution with far reaching implications for the whole basis of industry. It is the convergence of micro-electronics, computers and telecommunications. It encompasses control technology, systems management, information engineering, manipulation of data of all kinds, organisation of banks of knowledge, remote sensing, artificial intelligence and expert systems. It can extend and appeal to all human senses. It can be combined with other technologies to multiply their effects — with biotechnology, medicine, engineering and manufacturing production.
8. Industry must accept some responsibility for the negative aspects of its image — "dirty", "boring", profiteering", "rife with conflict", but there are grave social dangers if these images are perpetuated. Industry can do a great deal to project a *positive image* both by its actions and by providing links with education. Many examples are described in this book.
9. Ultimately the success of industry is dependent on *people*. The following are important:

   (i)   consensus, teamwork and leadership,
   (ii)   orientation to success,
   (iii) positive social environment,
   (iv)   innovation and enterprise,
   (v)   adaptability to change,
   (vi)   basic numeracy.

The Role of Tertiary Institutions in Development

Introduction

This section deals with the role of the tertiary sector in development. Williams takes up the points made by Ramachandra, and by Somerville and Sabberwal (see previous section) that the formal educational system is not producing in its graduates the attributes and skills required by industry. He puts forward suggestions for the revision of the university degree structure. Nevertheless, he sounds a warning that there is a considerable disparity between university and industry in terms of resources, and it is unlikely that there are many points of contact between them in many developing countries.

Sabberwal describes a scheme in the UK for collaboration between universities and industrial firms. Indeed, collaboration between industry and the universities in developed countries, in contrast with many developing countries, is being actively encouraged by industry and the academic institutions, as well as by governments. The reasons why this is not so prevalent in developing countries are complex, but in the contribution by Suryan and Murthy there is an example of a successful transfer of expertise from university to industry. Achi uses the concept Universities of Technology as a means of producing graduates in Nigeria who are immediately useful to industry, and as institutions which can work alongside scientists and technologists in industry in solving manufacturing problems. Powell then describes a Technology Consultancy Centre at the University of Science and Technology in Kumasi in Ghana, giving examples of the work done in transferring technology to small industries.

In many countries industrialists are also members of the academic faculty and come for short visits (a day or a week at a time for example) to collaborate in the teaching programmes. An example of a further stage is given by McIntyre and Rowe, both industrialists who forsook industry for academia and helped create a chemistry course in which much of the chemistry is taught in an industrial context, pinpointing the economic, technological and social implications of the chemical industry. McCormick also describes how these factors can be integrated into an undergraduate course.

It has been pointed out that industrial practice changes much faster than educational practice. So it is in tertiary education. Creating a new course, such as that described by McIntyre and Rowe is relatively rare. The later chapters in this section are concerned mainly with changes and additions to existing courses, and which take up relatively little time. However, they do have a significant effect on the educational process, for they are relevant and motivating — and are student-centred.

Sethu Rao describes the Student Project Programme in Karnataka in India, which encourages engineering undergraduates to undertake projects concerned with the basic needs of villagers. The students use local resources to complete their projects. Sane, also from India, describes how students are helping to develop and make low-cost, locally-produced equipment for science teaching in colleges and schools. Again self-reliance is the key to this work as local resources are used. Leisten from Malawi uses projects in his university chemistry course: he points out the value of this sort of teaching if industry does indeed need "chemists who can not only look after the routine tests, but can also sort out the problems that arise". Kulkarni and Witcoff describe in-service workshops in industrial chemistry for industrialists — and this surely is the sort of course which many would like to see available for schoolteachers as it is in the UK and the USA.

39

University-Industry Relationships in Developing Countries

H. J. WILLIAMS

Njala University College, Sierra Leone

Before independence industry in developing countries was for the most part financed and directed by companies based overseas and was geared almost entirely to exploiting the territories' mineral and agricultural wealth. Local participation was limited to the provision of labour at the lower levels and indigenous technical skills were effectively confined to the small-scale activities of the local artisan. With the coming of independence, therefore, the need for skilled manpower to replace the expatriate, to attain economic as well as political independence, became paramount. New institutions for secondary and tertiary education were founded throughout the Third World in the years immediately preceding and after independence, and the dominance of the foreign company was challenged by outright nationalisation or, more often, by renegotiation of trading terms to favour national rather than multi-national interests. State enterprises were set up to control and exploit national resources, and the indigenous entrepreneur was free to take advantage of widening opportunities in commerce and local industry.

Universities have been producing graduates in all disciplines for over 20 years. These bare facts might indicate a successful relationship between industry and the universities; closer examination, however, reveals serious problems of understanding and early resolution of these must be regarded as top priority.

Employment of university graduates in industry

The employment patterns of university science graduates in the Third World show disturbingly similar trends in many developing countries[1,2] with no less than 80% of first degree science graduates finding employment as teachers in the secondary schools and the remainder divided between those who go on to study for a higher degree before

229

returning (usually) to academic life and those who find work in Government laboratories or industry. Only about 10% of science educational output at the tertiary level therefore serves to meet the manpower needs of industry while most of the rest merely underpins a self perpetuating and, in terms of national development, ineffective educational recycling. Most graduates who find work become schoolteachers not so much because they prefer teaching as a career but because suitable jobs in industry are not available.

Why this should be so may be attributed to a distinct preference by industry in developing countries for training future managers and senior technicians from a lower educational base than that of the university graduate. Complaints that university graduates lack practical sense and the scientific knowledge expected of them, and that their attitudes towards working as members of a team rather than as individuals leave much to be desired, are widespread. Small secondary industry, too, cannot afford the salaries that graduates have come to expect. Many companies, both privately and publicly owned, clearly feel that young apprentices recruited from school are more suitable material for the type of employee they require than the university graduate who, although preferable for some positions, has to undergo further training in any case and may have inflated ideas regarding his place and importance.

This trend is particularly evident in very large corporations of long standing, such as the mining and the oil companies. In many countries "multinational" companies were established and functioning long before any form of tertiary education existed and clearly the emergent universities have been able to have little effect on the way these companies are run. In Zambia, for example, where copper mining is the mainstay of that country's economy the university has neither the expertise nor the facilities to influence the mining companies' scientific methods or recruitment policies in any way. In Sierra Leone, likewise, the diamond companies and the university operate in separate spheres with only the field of economics offering any real point of contact.

Although these may be extreme examples, the system of recruiting potential senior technical and managerial staff direct from school and sending them abroad for specialist training without reference to the national universities is still very common among multinational companies. Technical problems, too, tend to be referred back to the parent company overseas and little attempt is made to set up local development units or involve the universities in research of common interest. While such policies may be financially advantageous, at least as far as the company is concerned, they do nothing towards creating local employment or promoting the scientific ethic in the host countries from which their wealth is derived.

The university degree structure

In changing their academic structures to take into account the recruitment needs of industry, the universities are faced with enormous problems, not least to define exactly what is required. The generally low and uneven quality of the student intake means that most science degree programmes must be very broadly based to include remedial and core subjects such as mathematics and the language of instruction, while constant financial constraints require that the degree be completed in as short a time as possible. Under such circumstances the universities' room for manoeuvre is severely limited and only a radical reappraisal of the purposes of university education can hope to bring about fundamental change.

In the author's view the key to reform lies in the fact already pointed out that industry prefers to recruit apprentices with an educational background guaranteeing a firm base on which to build its training programmes but below the level of the full university graduate. To this end an under-graduate programme for science students divided into two distinct units, Parts I and II, is proposed. The essentials of this proposal are, briefly, as follows.

In Part I (2 years) the student would receive an all-round general education in science-based disciplines after which a graded diploma would be awarded to all participants. Part II (3 years) would comprise the degree course proper in specialised subjects and would be open only to students judged suitable from their performance in Part I. Part I would include all necessary foundation and remedial courses and would bring science university entrants to a uniform and thus more widely recognised educational standard. With proper consultation with all parties concerned the educational level achieved in Part I would serve as an acceptable basis for subsequent careers as recruits by industry, trainee teachers etc. as well as for students going on to study for Part II. Wastage due to student drop-out would be reduced to a minimum and there would be a greater efficiency in terms of college places available to students and consequently better use of the public funds needed to maintain them could be achieved. Students who clearly showed that they were not graduate material (and it is virtually impossible to forecast student performance at university on the basis of their school-leaving examinations) would leave at the end of Part I with their certificates of success rather than failure.

The universities and small-scale industry: applied research

As well as restructuring the degree programme changes in attitudes towards industry need to be brought about, not only in students but also in academic staff. Professors and lecturers should become acquainted with the problems of local industry, preferably by visiting factories and industrial sites where they could talk to employers, arrange student field trips and holiday jobs, and generally carry out a very necessary exercise in public relations. Employers could be invited to send representatives to faculty meetings and be encouraged to give their views on syllabus content and other relevant matters. All this requires good-will, persistence and commitment by all parties to achieve real results.

With regard to scientific research carried out in Third World universities much has been written and said on the irrelevance of "pure" research and the need for involvement in what is loosely termed "appropriate technology". A large number of Third World universities do in fact have research units committed to investigating and improving the technology underlying a wide range of small-scale local industries. The problem here, surprisingly perhaps, seems to be not so much the development of applied technology as its implementation[4]. The response to published results is said to be minimal and it seems that the universities must adopt a "hard sell" approach if they are to bring about real change in local industry's habits and attitudes. Failure to implement locally developed technology would inevitably mean continued dependence on imported skills and a reversion by discouraged academics to their traditional "ivory tower" activities. An example of a successful unit is given by Powell (p. 249).

The universities and large-scale industry

While useful co-operation between the universities and small-scale local industry would appear to depend for the most part on initiative by the universities, the relationship between the universities and large-scale established industry, for example, the multinationals, is very different. With the possible exception of some state controlled enterprises (which are effectively run on much the same lines as any large-scale concern), industry exists essentially to make profits. And, while heavy industry may feel it has no need of the universities to improve its profits, large industrial concerns in developing nations are certainly in a position to aid the universities in a number of ways.

Although some Third World universities were originally well endowed there are few which are now free from acute problems concerning technical and financial support. It is here that co-operation with industry would be of the greatest value, particularly with regard to the upkeep and effective use of expensive capital equipment. The case for mutual access to available instrumentation by all interested institutions in countries short of basic resources seems obvious. Other fields open to beneficial co-operation between the universities and industry include mutual consultation on the types of research topic in universities most relevant to local needs and a systematic evaluation by industry of such projects once they are set in motion. The possibilities of consultancy by university staff in industry should be examined further and would normally concern the scientific basis of an industrial problem. Consultants, however, must be experts in the field in which their advice is given — which underlines the case made elsewhere for not taking short cuts regarding underlying theory in degree courses.

Conclusions

The universities create the conditions for innovative research and the universities represent the scientific elite of their country. Industry, on the other hand, can supply the technical muscle which the universities lack and perhaps see more clearly the economic value of proposed research programmes. Proper consultation between the universities and industry on the lines discussed above is essential and if this is not forthcoming, or agreement cannot be reached, it is then up to Government to intervene and, if necessary, to chair proceedings. As elsewhere, government in developing countries must ensure that a proper balance is reached between the rewards due to the owners of industry on the one hand and what industry in turn contributes towards national development on the other. The poor nations of the world canot afford to do otherwise.

References

1. Proceedings of Unesco International Symposium on University-Industry Interactions, Toronto, Canada, December 1978.
2. "Aspects of University-Industry Interactions in Developing Nations", H. J. Williams, in Proceedings of Regional Conference on Chemistry in Africa, Nairobi, Kenya, 1980, pages 200–220.
3. "The Place of Chemistry in African Universities", H. J. Williams, *J. of Chemical Education*, **53**, 789, 1976.
4. "University Involvement in Appropriate Technology", W. Hamilton, in reference 2, above, 1980.

40
The Teaching Company Scheme

A. J. P. SABBERWAL
Turner and Newall Research, Rochdale, UK

The Teaching Company Scheme forms one of many initiatives in the United Kingdom, sponsored by the Science and Engineering Research Council (SERC) and the Department of Trade and industry to develop active partnerships between Universities and Polytechnics and manufacturing companies.

The main aims of the scheme are:

(1) to raise the level of manufacturing performance by effective use of academic knowledge and capacity;
(2) to improve manufacturing methods by the effective implementation of advanced technology;
(3) to train able graduates for careers in manufacturing;
(4) to develop and retrain existing company and academic staff;
(5) to give academic staff broad and direct associations with industry for research and as background for teaching.

The permanent academic staff of a university or polytechnic participating in the scheme are assisted by high calibre graduates recruited in consultation with the company for 2 year academic appointments as Teaching Company Associates. The Associates, normally based full-time at the company, work in collaboration with the company and the academic staff on tasks within the programme. In addition, the university or polytechnic arranges on-going tuition according to personal and programme needs. Associates may register for higher degrees, but essentially they aim to enter industry, either through the specific company or outside it.

The Science and Engineering Research Council and the Department of Industry make a grant to cover the basic salaries of the Associates and academic support costs. Sometimes further help is provided through a senior assistant who can relieve the academic staff from some of their teaching so that they can devote more time to the Teaching Company. A Management committee with senior industrial and academic membership supervises the operation of the scheme and the scheme is administered through the SERC by a Director and other supporting staff.

Expertise in the university is transferred to the company by the Associates who also bring back to the academic departments an understanding and knowledge of the problems that arise in day-to-day industrial life. In order for this arrangement to work at all appropriate levels of the company and the university, it is essential that good relations should be established at a senior (usually Managing Director/Professorial) level. Given the commitment at the top, middle and junior management together provide an environment in which the Associates can function effectively.

The benefits can be summarised as follows. *Companies* benefit from the contributions of the academics and Associates to their business objectives and from the identification and training of able and ambitious Associates for potential senior appointments. The *universities* and *polytechnics* are able to extend their research and postgraduate and post-experience teaching beyond the classroom and laboratory to manufacturing companies. Academic staff benefit by such close liaison with industry and continue the association through consultancies — even after the scheme terminates. The *Associates* learn by the experience of working with senior academic and industrial staff on demanding tasks with commercial, social and time factors. The *country* benefits from the short and long term application of academic resources to manufacturing and the demonstration of opportunities in industry for high calibre young men and women.

There are now about 140 programmes and a current budget of £5 million a year. It is hoped that it will rise to £11 million by 1988, allowing for about 200 programmes. Contributions from the participating companies, about 20% of the total costs, are expected to rise to at least 33%. Initial projects dealt with the "traditional" area of batch manufacture in mechanical engineering but recent ones include programmes in civil and chemical engineering, biotechnology, plastics processing and electronics.

Some examples

A Teaching Company programme contains three basic components:

(1) A company committed to a coherent set of significant improvements in its manufacturing practice;
(2) a university able and ready to contribute its knowledge to the implementation of such improvements;
(3) a number of able young Teaching Company Associates.

Some examples of successful Teaching Companies are described below.

Austin Rover/Warwick University

At Austin-Rover plant, an integrated manufacturing system was designed and installed in 1981 to control the manufacture of light commercial vans. The system is run from a minicomputer and the first schedules were implemented ten weeks after the project began.

Some of the achievements were that the stock holding of sheet steel was cut from 1700 tonnes to 400 tonnes; the total stocks were reduced from £4 million to £1 million; the non-standard vehicle construction time was reduced from 12 weeks to 3 weeks; better usage of materials, resulting in annual savings of £400,000.

John Williams of Cardiff plc/University of Wales Institute of Science and Technology

The John Williams foundry supplies a full range of castings to the major diesel and heavy automotive manufacturers. The Teaching Company programme is centred on reducing energy costs in the foundry, which consumes over 75% (£700,000 per annum) of the energy need in John Williams plc. Total savings of 18% (£135,000) have been achieved.

The measures taken include improved regulation of coke supplies to cupolas; improved ladle design, reducing the amount of alloy used; control of maximum electrical power demand, thus avoiding peak charges; improved furnace linings, the use of microprocessor control on an annealing furnace. The total cost of only one Teaching Company Associate was £24,000.

Other programmes

Bentley Textiles with the University of Aston on implementing new systems and technology in textile industry; Sperry Gyroscope with Bristol Polytechnic on reducing costs in the precision electronics engineering industry; Thames Board and DRG Markinch Mill with Imperial College, London on improving efficiency of production systems in the Paper and Board industry; Joseph Adamson with University of Manchester Institute of Science and Technology on computer aided techniques in the manufacture of pressure vessels.

There are also projects that deal with common problems for a number of small companies. Examples are Gauge and Toolmakers Association with Warwick University on computer techniques for one-off tooling in smaller companies; Small Manufacturing Industries Development Association with Hatfield Polytechnic on product development and other projects with a group of small firms.

Turner and Newall Project: T&N Materials Research Ltd (MRL) with the Department of Electronic and Electrical Engineering, Salford University

MRL is a member of the Turner and Newall Group of Companies, a

British based international manufacturing and mining group, which supplies materials, components and services to industry. Its principal activities include construction and industrial materials, automotive components, plastics and mining. The group devotes considerable effort to the development of new materials and processes and is a world leader in friction materials technology, in the manufacture of amino plastics, spiral wound gaskets, fibre reinforced materials and pipes. Although Turner and Newall's assets are widespread internationally, nearly 40% of its capital is employed in the United Kingdom. The group's sales turnover in 1983 was about £500 million.

MRL is the central technology development Company of the T&N Group. It is based in Rochdale, near Manchester and employs 45 people with an annual budget of £1 million. Its original term of reference was to develop asbestos substitute products. As this task was nearly accomplished, MRL was given new terms of reference to enable T&N to diversify into new business. MRL set itself the objective of developing businesses that are based on high technology and serve critical end uses.

As a result of a systematic search both in the UK and abroad, MRL concluded that surface treatment of materials by various techniques was potentially an attractive field to enter into, that the technique of ion implantation (in which high energy ions are bombarded at the surface of metallic components, are captured by the surface, leading to a modification of the surface), was a front runner and that United Kingdom institutions had a world lead in the technique. The advantages of this treatment are

Increased hardness, enhanced wear resistance, reduced friction, anti-galling/pick up properties, corrosion resistance, fretting resistance, fatigue resistance.

MRL had no expertise in ion implantation techniques, but the nearby University of Salford is a centre of excellence for the technique. MRL had a number of objectives in evaluation of this technique:

(a) it could achieve savings in its manufacturing of friction materials (disc pads, etc) by lengthening the lives of expensive compression dies and possibly also in the tools used for its extensive gasket cutting operations;
(b) it could develop new business, providing improved tools to manufacturing industry in the North West of England;
(c) it could develop more effective materials to be used as bearings, valve seats, stems etc.

The University of Salford, on the other hand, would have more resources through which it could increase its understanding of the new materials and the new techniques than it had hitherto.

A Teaching Company proposal was put forward and accepted for three Teaching Company Associates. They will broadly function in three areas:

(i)   marketing survey of current and potential base load and in-company needs;
(ii) providing a customer service;
(iii) developing newer materials and techniques.

Three academics from Salford also support the programme. The University agreed to site one of their machines (a new machine would cost in the range of £100,000 — £250,000) in the premises of a nearby Turner and Newall Company.

The Management Committee consists of three Professors, two Senior Managers from MRL and is chaired by the Managing Director of MRL (the author) and a representative of the Teaching Company Directorate.

The normal procedure is that the Management Committee meets every quarter and each of the Associates provides a report on his work during the period.

The Teaching Company became operational in June 1984. The budget for two years is £170,000, the industrial contribution being £55,000 and the Government contribution being £115,000.

41

Appropriate Research for Development

G. SURYAN and A. R. VASUDEVA MURTHY
Indian Institute of Science, Bangalore, India

One of the most serious deficiencies in many developing countries is the lack of co-operation between universities and industry. Far from having integrated and planned initiatives such as the Teaching Company Scheme described in the previous paper, there is often little contact, in particular in designing research programmes which can lead to a successful transfer of science and technology to industry.

This paper describes one project which has led to successful collaboration. The project is concerned with the production of high purity silicon, which has become an indispensable and critical material of great strategic importance to India. Currently we have to import all silicon from Germany, Japan and the United States and none of them have spare capacity. Thus supplies of electronic grade silicon are vulnerable to the vagaries of the commercial policies of the exporting countries. It is doubtful whether these countries will agree to the transfer of the technology to a developing country like India. The production of high purity silicon in the country has therefore become imperative. The policy of the Government of India of self-reliance assumes great importance in such a programme.

Over the last 20 years, we have been investigating at the Indian Institute of Science in Bangalore methods of preparation on a laboratory scale of silicon and silicon compounds using materials which are indigenous. No foreign exchange is involved in procuring raw materials such as ferrosilicon, metallurgical grade silicon, chlorine, methyl chloride, hydrogen chloride, hydrogen, industrial alcohol, etc. They are all available in tonnage quantities.

The technical know-how obtained in our laboratories has now been transferred to industry (Mettur Chemical and Industrial Corporation Ltd.) who, in turn, are using indigenous machinery and fabricating devices. The quality of the silicon and its compounds now produced conform to international specifications and the cost is competitive in the world market.

High purity *silicon tetrachloride* is being produced and Mettur Chemicals is one of the few companies in the world able to manufacture it. The manufacturing capacity is of the order of 600 to 1000 tons per year. As it is made in modular units, the production can easily be scaled up within a short period to meet growing demands. The machinery and the reaction units which are used by industry are all fabricated in our own workshops in the Institute.

Furthermore we are now self-sufficient in *ethyl silicate,* which is used in foundries and in the paint industry for anti-corrosive coatings, amongst other uses. Indeed Mettur Chemicals have already exported this compound to the United States and to Europe, thereby earning valuable foreign exchange.

The high purity silicon for the electronic industries is produced by the hydrogen reduction of silicon tetrachloride, trichlorosilane or a mixture of the two. Mettur Chemicals propose to manufacture about 25 tons of polysilicon and 10 tons of single crystals and wafers a year, all conforming to international standards. High purity quartz crystals are also made which have many industrial applications. The necessary methods have all been standardised at the Indian Institute, and the necessary equipment has been built locally for the use of industry.

42
How Best can the Universities of Technology help Industrialise Nigeria

P. B. U. ACHI

Federal University of Technology, Owerri, Nigeria

Technology can be considered to be, as defined in the Encyclopaedia Britannica, "the industrial processes that replaced craft operations". Today therefore, technology can be defined as that branch of engineering which concerns itself mainly with the implementations of the findings in engineering. Engineering on the other hand organises facts and findings from the sciences and applies these to the formulation and implementation of solutions to physical, environmental and general problems related to human needs.

Technological education for industrial applications

In Nigeria since the oil boom in the 1970s the Federal and State Governments have looked for ways to stimulate the growth of design and production industries engaged in stable local manufacturing. There are at least six vehicle and other assembly industries recently established as partnership ventures between the governments and some foreign companies, although these industries only assemble completely knocked down components (CKDs) while the design and manufacture values are added in foreign countries.

The problem today is that the technologies are not really being transferred and today the country is faced with the dilemma of scarce foreign exchange to purchase CKDs and consumer goods. Therefore the governments are now emphasising indigenous technology as priority in their industrialisation policies.

The establishment of Universities of Technology can be seen in this context, in which advanced technical education is obtained which is significantly different from that imparted in the existing higher institutions. Since the seventeen conventional universities in Nigeria (as opposed to Universities of Technology) are most likely to continue their traditional methods in liberal, scientific and technical education, the eight new

Universities of Technology are expected to be complementary to and not competitive with these conventional universities in curriculum design.

In each of the conventional universities there exists at least one faculty or school for the pure sciences while only six of the universities have established reasonably equipped and staffed faculties for engineering (or technology) each capable of graduating an average of 100 candidates every year. The curriculum leading to the first engineering degrees awarded in these faculties consists of the traditional engineering science subjects (almost entirely theoretical). The mechanical engineering curriculum, for example, consists of such applied sciences, like mechanics of solids and fluids, thermodynamics and its theories in thermal equipment, theoretical design of machines, engineering mathematics, mathematical analysis of control systems, textbook-based general engineering economy, classroom-based engineering drawing and minimal laboratory and workshop practices.

This form of technical education may be theoretically sound but seriously lacking in the practical aspects. As a result, on first entry into a job many employers find these new graduates practically unproductive. So the company has to incur the extra costs to retrain them often for a minimum of 2 years in industry before they can be given full responsibilities.

The new Universities of Technology were established to conduct technical research and to impart practical skills (among other things) to its new graduates; they are expected to bring about some form of technological and industrial revolution in Nigerian society by producing graduates whose engineering education is geared to industrial applications. These graduates are expected to fit into industry much more quickly than those who have received a more general engineering training, becoming productive even while receiving the necessary orientation for specific skills peculiar to the industry. What Nigeria requires now is qualitative as opposed to quantitative education.

The seventeen existing (conventional) universities, over twenty polytechnics, and the eight new Universities of Technology can be more easily seen to serve complementary purposes if the latter were to draw the bulk of their intakes from the graduates and diplomates of the former. Diplomas, bachelors or masters degrees would then be the entry qualifications to all departments in a University of Technology. Thus, these new universities will be used to retrain some of the graduates from the conventional institutions where they have been theoretically prepared. When a University of Technology is planned this way in Nigeria it will be freed from the distractions and enormous costs of basic undergraduate education and concentrate meaningfully on practice-oriented engineering and technology. The University is then reasonably equipped to take on profitable research contracts from within and outside the country, offer more reliable consultancy services, and engage in competitive manufacturing businesses using industrial centres which should be established within the campuses. At the present time this is not the case in the new Universities of Technology though the idea of industrial centres within the campuses is being gradually accepted.

The centres for industrial studies: transfer for industrial development

In Nigeria (and indeed Black Africa) the development of an "indigenous" technology must not be considered to be "re-inventing the wheel", as it were, by going back independently to the European Industrial Revolution. We shall acquire the technology of the day by transferring it into specific centres of technical education from which the knowledge will be spread to Nigerian industry.

The future of developing countries with *temporary* sources of revenues like petroleum can be seen to depend on their ability today to establish profitable industrial systems geared to competitive production with home and specialised export markets in mind. A University of Technology in a developing country should, therefore, be principally oriented *ab initio* towards the training of indigenous manpower biased towards the production of those goods and services which are often scarce in these countries. The pre-occupation of these universities should be the practical research and education in the art of profitable design and production of capital and consumer goods because the overriding responsibility of an industry is the **profitable** production of goods and services.

Centres for industrial studies (now in the pipeline) which should be multi-disciplinary in nature can be established in each University of Technology to serve the practical objectives of an industry while the academic departments should be principally engaged in technical and managerial industrial research and prototype development. The industrial centres and the academic departments should be engaged in imparting education in:

the technology of machines and manufacturing processes; product conceptualisation and design, development of local industrial raw materials; the design of profitable production systems; the planning and control techniques for the production of commodities; quality control techniques for all ranges of local industrial products; the use of computers for the efficient production of goods, services and maintenance; the development of superior species in agriculture and the processing of agricultural products.

The industrial development centre will liaise with all the disciplines in the university, local and foreign industries to impart the specific initial skills in technology and management to the students of the University of

Technology. The centres will then be the beehive of technology transfers between Nigerian (or any other developing country) and foreign companies from the industrialised world.

Technology transfer has become a very controversial phase in Nigeria today as a result of the observed reluctance of the industrial nations to transfer their technology to developing countries, although some economic leverage can negotiate some transfer, for example in motor-cycle design and manufacture, brewing plant design, etc. The centre for industrial studies when established in the Universities of Technology can be used as nerve centres for the acquisition by Nigeria of technology from other countries. One of the ways will be by official technology transfer agreements between the governments and universities on the one hand and foreign companies with trade ties with Nigeria on the other. Supervised training visits between the industrial centres and the parent companies abroad could be arranged for Nigerian students and staff from the selling nations. These trainings would be expected to last from a few months to a few years depending on the technology envisaged. The students in the University of Technology should spend more of their time in the industrial centres than in the classroom. While in the centres the students would be fully involved in all the stages of the research development, production and maybe marketing of products which would be periodically selected and later handed over to the industries.

Nigerians with various specialised technological skills still residing abroad should also be given enough incentives to attract them to the various centres for industrial studies. Nigeria should also encourage foreign nationals with proven technological expertise to seek Nigerian nationality. Moreover, all those Nigerians (indeed Africans) at home with technological achievements should be mobilised into the industrial centres. In all technology transfer ventures the expertise within the universities and the local industries should be allowed to play the central roles in negotiations and implementations.

Conclusions

Today the existing industrial culture in developing countries allows practice-oriented technological higher education as a viable proposition. The new Nigerian Universities of Technology can be used to bridge the technical gap between the existing conventional universities and industry by imparting a good degree of the specific skill needed by industry. This they can do most effectively by recruiting postgraduate students from the graduates of management, science and engineering from the conventional universities. Thus the Universities of Technology can be seen to be complementary to the existing universities and industry. The Centres for Industrial Studies in the new universities will work with all disciplines of the University and the companies involved in the acquisition of foreign technology to establish profitable basis for our industrial development.

43
Technology Transfer to Small Industries in Ghana

J. W. POWELL

Technology Consultancy Centre, University of Science and Technology, Kumasi, Ghana

Technology transfer and economic development

While most thinking men agree that the problem of poverty in the Third World can be solved by the transfer of technology, there have been, and to some extent remain, several schools of thought concerning how it is to be best achieved. For many years it was believed that the large-scale advanced-technology industries, developed over decades, and even centuries, in the industrialised countries, could be transferred intact and unmodified into a pre-industrial environment. The type of crash programme in industrialisation which this approach engendered was experienced in Ghana during the Nkrumah era (1957–66). In Ghana, as elsewhere, it failed, and the reasons for its failure have been the subject of detailed study and copious documentation. It is now more generally appreciated, especially following the writings of Dr E. F. Schumacher of "Small is Beautiful" fame, that to succeed, an industrialisation process should grow, as all natural organisms grow, slowly but steadily from the grass roots.

The university and the informal sector

The University of Science and Technology, Kumasi, was founded by Kwame Nkrumah to train the high-level technologists and managers for his "big push" towards industrialisation. It was, therefore, modelled on Western institutions designed to feed advanced technology industries and, in common with many universities elsewhere, little attempt was made to adapt it to the African environment in which it was destined to operate. However, following the overthrow of Nkrumah and the collapse of his industrialisation programme the University was forced to re-examine its role and attempt to adapt itself to the new situation. This process began slowly and somewhat painfully in the late 1960s and gathered momentum and purpose during the decade of the 70s.

The metamorphosis started in 1968 when approximately 40 lecturers representing several faculties of the University formed themselves into the Kumasi Technology Group to offer free or low-cost consulting services to small-scale and informal industries in Kumasi. This group flourished for only 2 years but its role was taken up again in 1971 by the Suame Product Development Group. Suame Magazine is the largest informal industrial area in Ghana. Within its extensive confines are gathered some 27,000 craftsmen, wayside auto-fitters and apprentices in an estimated 2,000 shanty-style workshops. To many the Magazine may represent a scene of squalor and disarray but to those of keener sight it can be seen to demonstrate a very considerable potential in terms of human skill, ingenuity, enterprise and the will to succeed in a difficult and sometimes hostile environment. In associating itself with Suame Magazine the University was turning to explore the grass roots of indigenous industrial-isation in Ghana. This was not a move which, in 1971, was likely to gain official approval. The inhabitants of the Magazine were squatters whom the authorities had moved on once and were attempting to move on again. There was little public acknowledgement of the services which the Magazine provided to the community at large. However, it is the duty of a university to probe ahead of the conventional wisdom and the measure of its success was indicated by the fact that by the close of the decade the Government of Ghana had been persuaded to finance a project to establish an Intermediate Technology Transfer Unit operated by the University of Science and Technology in the heart of Suame Magazine. But this is to jump ahead of our narrative.

The interest on the part of some academic staff in the problems of informal industries coincided in the early 1970s with a demand that consulting for outside bodies be recognised by the university authorities as a legitimate means of earning additional income. After some deliberation and consultation with outside bodies the University formalised both its interest in grass roots development and its channel for providing consulting services when it established the Technology Consultancy Centre (TCC) on 11 January 1972.

The Technology Consultancy Centre was, and is, exactly what its name implies. However, it is a centre dedicated to the solution of all the technical problems brought to it by the citizens of Ghana regardless, in the first instance, of ability to pay. In some years up to 20% of its effort has been expended on behalf of governmental or large business interests which pay conventional consulting fees. However the major portion of its effort has always been concerned with small-scale, informal, craft industry and agricultural projects for which funding is sought on a largely piecemeal basis from sources within Ghana and overseas. In particular, the work of the TCC has been strongly supported by non-governmental organisations in Britain, Canada and the USA. A measure of the demand for its services in Ghana and the support of its work from overseas is indicated by the fact that the annual budget of the TCC grew from $4,000 in 1972 to $400,000 in 1983. In 1983, over 99% of all expenditure was on small-scale industry development projects.

The Technology Consultancy Centre

The Technology Consultancy Centre is established as a semi-autonomous research centre within the University of Science and Technology. It is managed by a Management Committee, chaired by the Vice-Chancellor and including all Deans of Faculties, which meets three or four times a year. Day-to-day affairs are in the hands of the Director who reports directly to the Vice-Chancellor. The Centre is therefore independent of the academic faculties and can draw upon their services in accordance with the needs of individual projects. The TCC operates its own bank account and employs its own accounting staff. It operates its own mechanical engineering workshops and operates and maintains its own fleet of vehicles. These resources enable it to react flexibly and effectively in response to the problems that are brought to its attention. In 1983, the TCC employed 13 professional staff and a total staff of 70 of whom 31 were established with the University and the remainder were casually employed on individual projects.

The key role within the TCC is played by the Project Officers of whom there are presently ten. A Project Officer makes the first contact with the client either at the TCC office or at the client's place of business. His first task is to determine if the client is serious in his intent to develop his activities. Usually the client is tested by being given some task to perform in his own interest. This may consist of undergoing training at one of the TCC's pilot plants, conducting a simple marketing exercise for his new product or gathering information on the cost and availability of materials obtainable from local sources. The Project Officer makes it clear to the client from the outset that to work with the TCC is to enter into a partnership in which he will be required to make a serious effort on his own behalf and in which the help given will be matched to his own efforts. In this way scarce resources are husbanded and brought to bear in greatest force in support of those with more apparent entrepreneurial abilities.

Once the Project Officer is convinced that he has a serious client on his hands he becomes the co-ordinator of activities designed to solve the client's technical problems and related managerial and financial problems. In many cases technical advice is sought from a consultant in one of the University's academic departments. The Project Officer ensures that the advice is followed by the client and feeds back to the consultant any further problems that arise. He ensures that laboratory tests are carried out or that equipment is made to the consultant's direction at the TCC workshop. He may arrange visits by the consultant to the client's workshop or the transport of plant and equipment to any part of Ghana. He may continue to advise the client over a period of months or years and will assist him with feasibility studies and cash flow projections when the client needs to raise finance from his bank. At some point the Project Officer will write a Case Study of the project, especially when it is the first of its kind. The Case Study then serves to guide future projects of a similar nature and, as all successful projects tend to be repeated many times, the Project Officers often become sufficiently experienced to advise clients without recourse to a consultant.

Many projects brought to the TCC have resulted in the establishment of pilot production units. In the early days, pilot plants were established on the University campus and managed by TCC and faculty staff. These made such commodities as soap and caustic soda, steel bolts and nuts, hand woven textiles, low-cost building materials, well pumps and small-scale process plants. The objectives of these production units were:

(a) to train craftsmen and managers in the skills of the new industry;
(b) to complete product development under production conditions (adapting the product to the needs of the market and available production technology);
(c) to test the market for the new product in a realistic way, and hopefully,
(d) to demonstrate to entrepreneurs the viable operation of a new industrial activity.

The pilot production units support the work of the Project Officers by demonstrating production processes, providing on-the-job training, supplying reliable costing data, exploring raw material supplies and proving the market. The entrepreneur is by definition a person who takes risks in pursuit of ambition. The existence of a pilot production unit which is operating viably is an effective means of reducing the real and apparent risks to be run by the entrepreneur and thus of encouraging more to enter a new field of venture. Clients who would remain unconvinced by the most thorough paper study are often persuaded when they can see and take part in the production process and examine the accounting records of actual manufacturing and trading experience. In this connection it must be borne in mind that small-scale and informal industrialists are often illiterate and almost always unaccustomed to accepting guidance through the medium of the written word. Thus the pilot production unit is one example of how the technological university adapts itself to the needs of its new clientele. Where research reports cannot be relied upon to effect the transfer of technology, the results and development activities must be translated into practical examples which all can appreciate irrespective of educational background or attainment. In conceiving the establishment of pilot production units the University took its first step towards transferring technology to the informal industrial sector. The second step involved the establishment of Intermediate Technology Transfer Units (ITTUs).

Although much was achieved, and even more was learned from the operation of on-campus production units, it began to be realised that their effectiveness as a medium of technology transfer was limited because their activities were not sufficiently exposed to the informal industrialists who might benefit from them. This clientele does not naturally relate to a university. Their education seldom extends beyond middle school level and very few have any business which brings them to the university campus. Some came to the campus to purchase the products of the production units to use in their own workshops but, apart from this, contact was only made by means of visits by Project Officers to informal industrial areas. The TCC realised that the activities of production units could affect the lives of far more craftsmen and entrepreneurs if the units were themselves situated in the informal industrial areas with the Project Officers working out of them. Thus evolved the concept of the Intermediate Technology Transfer Unit.

Intermediate Technology Transfer Unit (ITTU)

At the present time (1984) the Technology Consultancy Centre has for 4 years operated its first ITTU in the heart of Suame Magazine, Kumasi. A second one is also being established at Tamale in the north of Ghana. Each ITTU consists of a group of four to six production units each demonstrating the manufacture of a new product or the operation of a new manufacturing process. The aim is to keep the level of technology used in the ITTU one step in advance of that found in the surrounding workshops. It exposes the craftsmen to techniques which they can understand and learn and which they have some hope of acquiring for themselves. When the technology is transferred and the market can be satisfied from outside, the ITTU moves on to a new activity or a more advanced level of technology in the same line of activity. New activities continue to be selected from amongst those requested of the ITTU by clients.

When the client brings a new problem, or takes a serious interest in an activity already underway in the ITTU, the aim is always to attempt to establish the productive activity in the client's own workshop at the earliest possible opportunity. In pursuit of this end many services can be made available. The client or his employees can be employed in the production unit until they are proficient in the skills of the new industry. Help can be given in the raising of finance and in the purchase of machines and equipment. The ITTU has a pool of imported used machine tools which can be hired, loaned or sold to clients. Other machines and equipment can be manufactured in the ITTU for sale to the client. To encourage the growth of the business, the ITTU may sub-contract orders to the client or hold stocks of raw materials on his behalf. It is important to realise that informal industrialists do not often produce for stock or hold stocks of raw materials against future orders. They usually wait for orders to be brought to them and then demand an advance payment for the purchase of materials. As a consequence, their productivity and machine utilisation is very low and many business opportunities are lost. The ITTU has an important task to perform in changing the attitudes of its clients. This is one example of where the transfer of technology has a socio-psychological dimension which cannot be ignored. It is also a further justification for the involvement of a university at this level because many problems are of a multi-disciplinary nature and require solution by a group of experts drawn from several diverse disciplines.

University involvement

Some people may seriously doubt whether a university should involve itself with intermediate technology at grass roots level. While admitting that the task must be performed by some agency or institution, they doubt whether a university is the proper one. In this respect it is appropriate to point out that the problem of technology transfer at grass roots level in Africa is one which at the present time remains largely unsolved.

Unsolved problems are the proper concerns of universities. In getting involved in this work the University is evolving an effective mechanism for the transfer of technology which, in the future, can be employed by other agencies. The Government of Ghana is setting up a third ITTU at Tema and once expressed a desire to establish a total of nine in various parts of the country. If this network comes into being it will be administered by the Ministry of Industries and by Regional Development Corporations. The University will then probably revert to a supporting role; developing new products and technologies and providing training for Project Officers. However, this stage is still some years away and, in the meantime, much remains for the University to do in carrying its plans into action and further refining the services which an ITTU can and should offer.

Technologies transferred by the TCC

The means described have been used by the TCC and by the Suame ITTU to transfer a number of technologies. A group of twelve engineering workshops has been established producing a wide range of products including steel bolts and nuts, gear wheels, agricultural tools, carpenters saw benches and wood turning lathes, soap making plant, palm oil mill plant, food processing machines and charcoal kilns. Some 30 small-scale soap plants have been established and a similar number of rural palm oil mills. Beekeeping has been introduced and some hundreds of beekeepers have been trained and supplied with hives and other equipment. Farmers have been introduced to minimum tillage farming on the IITA pattern and some sixty farmers cultivating about 100 hectares have joined the programme. A pilot fish farm has been established and a research and training programme started which focuses on the husbanding of local species of tilapia.

Overall the TCC's work has created employment for several thousand people and has generated economic activity amounting to several millions of cedis annually. Thus it might be claimed that Ghana's only University of Science and Technology has gone a long way in adapting itself to the needs of its environment and has demonstrated the effectiveness of its new role. This role is somewhat different from that conceived by its illustrious founder. It is to be hoped that it is no less significant for the economic development of Ghana.

44

A University Chemistry Course Incorporating Industrial Economics and Technology

J. McINTYRE and D. J. M. ROWE
University of York, UK

Traditional chemistry courses in universities concentrate on chemistry as an academic discipline with a bias towards research, the area of major interest to the academic staff who design and teach the courses. But there are many areas outside research in which graduate chemists may have legitimate interests — development, production, marketing and environmental control to name a few. The authors of this paper, two industrial chemists with a range of experience in these fields, joined the University of York in 1973 specially to develop and teach the industrial content of a course entitled Chemistry Course 2.

British universities and polytechnics have long offered courses in chemical engineering and various specialised technological aspects of chemistry, for example ceramics, colour chemistry, and fuel science. There are also combined courses such as Chemistry with Economics, or Chemistry and Management Studies, in which the business component is not integrated with chemistry but is taught jointly to students of chemistry, engineering and other subjects. The distinctive feature of York's Chemistry Course 2 is that it is an honours chemistry course in which the technological and economic component is taught, with special emphasis on relevance to chemical industry, within the Chemistry Department so as to integrate the technology and the economics with chemistry in a way that would be unlikely to happen with service courses in other departments. All illustrations used have chemical backgrounds and there is substantial synergism between the chemistry, technology and economics. We believe a course of this type has many advantages. The objective is to produce graduates who (a) possess a corpus of knowledge appropriate to a professional chemist but having also an appreciation of the technological, economic and environmental constraints associated with the exploitation of chemistry; (b) have some knowledge of methods that can be used to assist in the solution of problems in applying chemistry to practical ends; (c) have the ability to collaborate with people from other disciplines.

Structure of the course

Chemistry degree courses at York, as at the majority of British universities, extend over 3 years of three terms each. At York the courses are split into two parts. Part I is common to all chemistry students and provides a basis in physical, inorganic and organic chemistry with some biochemistry and relevant elements of mathematics and physics. In Part II all the students spend about half of their time on a core of more advanced chemistry. Whereas Course 1 students devote the other half of their time to more specialised topics in chemistry and selected interface areas between chemistry and other sciences, Course 2 students are introduced to technological and economic aspects of chemistry.

Technological and economic content

Two major differences between industrial chemistry and chemistry in the laboratory are SCALE and ECONOMIC OBJECTIVE. These have important consequences.

Typically industrial processes are operated on a very much larger scale than in the laboratory, frequently at pressures above atmospheric, and often continuously. Except in small industrial units glass is not, as in the laboratory, a satisfactory material for construction of equipment, and the industrial use of metals raises the possibility of corrosion. The larger scale means that the surface to volume ratio is very much reduced so that heat transfer — seldom a problem in the laboratory either for heating or cooling — is a matter for serious consideration. Transfer of materials is usually simple and quick in the laboratory so special methods are rarely required but industrially provision has to be made for handling gases, liquids and solids. In the laboratory, inefficiency in the use of materials can usually be tolerated but industrially material efficiency is important economically and also because of the environmental problem of waste disposal. Ideally, therefore, industrial processes have to be intrinsically efficient, or alternatively unused reactants must be separated and reprocessed, and productive use found for byproducts. The larger scale in industry also multiplies the potential hazards of flammable and toxic materials, thus much attention has to be given to safety both in the design and operation of plants. In the laboratory, control of a reaction is usually simple and little instrumentation beyond a thermometer is required; on the industrial scale instrumentation is vital, control depending essentially on inferences from measurements of pressures, temperatures, flow rates and levels (with some analytical back-up). Partly for reasons of safety and partly for ease of control, steam is the preferred heating medium in plants but there is an effective limit of about 470 K and process conditions are often adjusted to enable steam to be used. Energy efficiency is seldom, if ever, a matter for concern in a laboratory preparation but heat (and sometimes pressure) recovery can be crucial to industrial success. These are some of the ideas that are rarely appreciated by the graduate from a conventional chemistry course but which we are trying to teach our Course 2 students through a survey of some of the basic elements of chemical engineering — chemical processing equipment, mass and heat balances, principles of fluid flow and heat transfer, instrumentation and instrument control theory, effluent control technology, etc.

Economic objective is usually insignificant in the laboratory; in industry it is all-important. Industry must be profitable to survive and still more so to thrive. Profit, a measure of efficiency, is basically revenue less costs so both of these are considered. Early in the course, in parallel with technological input, attention is given to costs in the chemical industry — in particular the nature of costs, variation of cost with output in single-product plants, costs in joint production, costs in multi-product plants, budgetary control and economies of scale. Then the revenue side leads to a consideration of market economics — perfect markets and markets with varying degrees of imperfection typically encountered in the chemical industry such as monopoly, certain types of oligopoly, and price discrimination. Logical continuations are marketing (with special reference to the chemical industry), forecasting, distribution and location of industry. Company accounts also figure.

At various stages there is integration between the technology and economics: an important area of overlap concerns new plants. Consideration is given to industrial research, scaling-up from laboratory scale through semi-technical and pilot plant scale to commercial scale, project planning and evaluation (with sensitivity and risk analysis), project execution and plant commissioning.

One important lesson that we try to get over to the students is that industrial activity does not just happen: it is people who get things done. This applies in all areas — research, development, design, plant operation (even in highly automated plants), marketing, distribution and so on. Moreover people of many different disciplines are commonly involved — chemists; chemical, mechanical and electrical engineers; accountants; and not least, operatives of various types. Considerable emphasis is therefore placed on the importance of communication. Also, managers operate in a legal environment and have to pay proper respect to legislation governing employment, safety and environmental protection which, therefore, receive due attention. Some management theory is incorporated too.

The dynamic nature of the chemical industry is constantly stressed. Continually we try to show how the industry evolves in response to the ever-changing technological and economic environment, hopefully anticipating events rather than just reacting to them.

Course methods

Over the four terms 5, 6, 7 and 8 (term 9 being left for revision) 140 lectures are given in technology and economics and another 140 in chemistry, the 280 in total representing about 8 lectures per week. As for chemistry, the lectures are backed up with tutorials in which students in groups of three or four meet their tutors for discussion of problems after having done some preparatory work: there is a tutorial each week and 16 in total are devoted to technology and economics.

In addition to the tutorials there are the Course 2 "workshops", usually of 3 hours' duration but sometimes lasting all day (with appropriate breaks). In total over the four teaching terms 120 hours are devoted to these workshops. In them students in varying groups of two to six depending upon the nature of the exercise tackle problems of a technological or economic nature or both. The problems invariably relate to chemical industry and in these exercises, which are made as realistic as possible within the time available, the students can work together, discussing suitable approaches and the information they need to seek, sharing the effort in calculations, pooling their experience, and finally presenting their conclusions: the problems are commonly open-ended and various groups can legitimately come to different conclusions in matters of design, marketing strategy etc. — there are not necessarily any "right" answers although there may be wrong ones. This in itself is a valuable educational experience as in most chemical problems students are presented with precisely the information that they need and can only draw a single right conclusion: real life is just not like that! Reference 1 provides an example of a study which incorporates chemistry, technology and economics and which is used in a Course 2 workshop.

Valuable as teaching within the University may be, it would be nothing without some direct experience of chemical industry. In this respect we are fortunate to have some major concentrations of chemical industry as well as smaller units within 100 km of York. Our relations with industry are good and we are able to take our students to a spectrum of chemical factories. The cumulative experience over a series of seven visits is exceptionally valuable. Each visit is designed, in collaboration with the factory management, to concentrate on some particular topic such as distillation, solid–liquid separation, instrumentation, scale-up, mineral processing, production planning, etc. but much else emerges on the day. On the journey out the opportunity is taken to outline the regional setting and resources as well as the general nature of the factory and its location; at the factory much can be observed of management styles, industrial relations, safety policies and procedures and the like. In many cases senior managers spend time with the students, often informally over lunch: these contacts, and the more formal question sessions at the end are most valuable in giving the students insight into varied aspects of industrial life.

Use is made of the experience gained on works visits in tutorials and workshops, while frequent reference is made in lectures to features seen or to be seen on such visits. These visits form an integral (and examinable) part of the course.

Conclusion

At the end of the course we have graduates with some knowledge of chemical industry and, especially, a useful awareness of what industry is about. We believe that they are better able than traditionally educated chemists to develop professionally from the outset of their careers.

Reference

1. Rowe, D. J. M. *Industrial Chemistry Bulletin,* **2**(2), 33, 1983.

45

Chemical Industry, Chemistry Teaching

D. McCORMICK

Manchester Polytechnic, UK

The fact that chemistry education and chemical industry are symbiotic should perhaps be more evident in Britain than many countries. However, the relationships between chemistry educators and industrialists are not always self-evident and harmonious. At countless comings-together of educationalists and industrialists, mutual recriminations fill the air. The industrialists' case can be summarised thus: chemistry teachers are not interested in chemical industry, only in academic chemistry; industry needs well-rounded flexible employees, able to contend with messy interdisciplinary "real world" problems, which are only partly scientific. Even a specialist in research and development areas (and these are in the minority) needs to appreciate the wider objectives of his employer. The products of chemistry education know little about chemical industry, are often reluctant to work in it, and are not particularly good generalists. In response to the industrialists' complaints, educators argue that they are given the responsibility of educating people who have expressed a desire to become professional chemists. This requires them to initiate students into an elaborate and ever-expanding knowledge system. This in turn necessitates teaching not only the facts, relationships and theories but also how knowledge is advanced in the discipline. Demands on time are considerable, and this is all that can be done; education does not stop with the award of a degree; teachers and researchers require further education and training — and so do those employed in industry; industry should not expect education to do its job for it.

These opposing views might appear to be irreconcilable. However the content of chemistry degree courses is of necessity eclectic; no two degree courses are identical. This suggests that material can be omitted without jeopardising the validity of a course. If other material can be shown to be beneficial both in general and vocational terms, then the case for its inclusion must surely be strong. As members of a relatively young institution, we at Manchester Polytechnic are obviously concerned that our students should receive a standard of education which is equivalent to that obtainable elsewhere. However, this has not prevented us from attempting to develop courses which respond to the new thinking about the functions of chemistry education. Two such attempts might be useful for the consideration of those who are seeking courses with broader aims than are usual. The first example are the courses which have been developed on the theme of chemical industry. These courses endeavour, by a consideration of the UK chemical industry in a national and international context, to give the student insight into the economic and other factors which govern the growth, location and organisation of a chemical firm, into the technical and economic factors which determine selection of chemical processes, and into the theoretical and practical considerations in the design and operation of a chemical plant. In teaching the above, an emphasis is placed on the changes occurring in the industry and the technological, socio-economic and legal factors which cause them. Case studies on the manufacture and subsequent application of groups of chemical products are used to illustrate amongst other things, the role of the applied research and technical service departments in industry. Regular reading of the industry trade journals is an accepted part of the course, and student assignments and a seminar paper are related to this. Various texts are used for formal reference.[1] Invited speakers from local industry are used in support of our own staff and works visits supplement laboratory sessions.

Some typical examination questions for the courses are given below

— Describe the way in which the sources of sulphur for the manufacture of sulphuric acid in the UK have varied over the last 20 years. Include in your answer an outline account of how each material is converted to SO_2 and also give the chemical, environmental and/or economic reasons for its changing popularity.

— Discuss the advances which have been made in chlorine cell technology during the last 15 years which have allowed the design and operation of bipolar, "plate and frame" membrane cells with DSA's. What are the advantages of such cells over conventional mercury and diaphragm cell installations?

— Expert opinion is divided as never before on the likely medium term future (15–20 years) for the Western European chemical industry. Outline and comment on the factors which give cause for such widely diverging views.

From the first two questions, it can be seen that while keeping a strong foothold in chemistry, the course material takes the student into a dynamic technological context. The students have relatively little difficulty in making this transition. In contrast, the material which predicates the last question — a resource pack of authoritative statements made by industrialists, economists, etc. — can be for some students most disturbing.

Nonetheless, most of the students, with practice, master the requisite skills.

The courses are not without their staffing problems. We have found, as might be expected, that a team-teaching approach is necessary. This is not simply a matter of assembling the appropriate expertise (some of which, incidentally, comes from outside the chemistry department). Each member of the team must fully agree with the aims of the course and his role in achieving them. Fundamental questions arise. To what extent should the course be more critical? Some think it should not: they maintain discussions degenerate into ideological slanging matches. Others say it should: man, not the technology, has to be master. The choice of futures is as important a subject as the choice of routes to chemical products. Perhaps this debate can be looked on as fulfilling a general education function for the lecturers!

A second area which we feel has the ability to fulfil some of the aims is Analytical Chemistry. In Britain, in the past, analytical chemistry has been the poor relation (particularly in universities) of degree courses in chemistry, often being taught in other branches of the subject. There is increasing concern about this primarily because industry is somewhat critical of the standard of education and training of applicants for posts in analytical chemistry. With this in mind a recent report suggests a syllabus for an analytical chemistry course which might be taken by all undergraduates in chemistry.[2] The syllabus proposed embraces the classical and instrumental methods and an introduction to general analytical procedures. As it stands, it would afford the student a considerable body of knowledge which he would not otherwise possess; however, it does not take into account changes in the thinking about the teaching of analytical chemistry which have been taking place for more than a decade. In the real world, "the principal focus of chemical analysis is the problem to be resolved rather than some sample related to it".[3] The teaching of analytical chemistry should reflect this: it should give the student the appropriate preparation for, and experience in, "real world" problem-solving. This thinking has led us at Manchester Polytechnic to design courses which put some stress on the analytical problem — problem definition, method selection and implementation, measurement, data-processing and evaluation. The courses in no way neglect the generally accepted essentials of analytical chemistry. On the contrary, by stressing the problem, the underlying principles and the methods and procedures start to be seen as essentials. When a degree of emphasis is placed on the use to which information will be put, the data which will yield this information, the requirements with regard to accuracy, precision and sensitivity, the nature, size and number of samples, and constraints such as cost and time, then analytical chemistry is, simultaneously, both "real world" and academic.

When the students are ready for it, often sooner than one might imagine, they are given real analytical problems to solve. The students work in groups of three or four, and in the first instance discuss the problem with the "client" (usually the lecturer in charge but sometimes another student who has brought a problem in from industry). They start to define the problem. They are prompted to ask questions: it pays dividends. What begins as the separation of closely related indoles in an industrial formulation turns out to be the quantitative determination of the main component — a different problem. They recognise that the request for a quick method for catecholamines in urine is really a request for a rapid clean-up procedure. This becomes the problem. The groups work independently, utilising whatever expertise and resources they think appropriate. They are given about 12 hours (spread over four weeks) to resolve their problem. As often as not, the problem remains unresolved at the end of this time. However, provided the students demonstrate that they have properly grasped the problem, have attempted reasonable approaches to it, and have understood the factor which delayed resolution of the problem, then they incur no penalty. Why should they? Real problems sometimes take months to resolve. It is only foolproof laboratory exercises which can be concluded in half a day.

What do the students gain from this experience? Since there are no laboratory sheets, they are forced to exhibit initiative. They become actively engaged in experimentation of their own design; cooking the results has no place here. They certainly learn how to co-operate, something which we, along with others, have discovered is not really engendered in traditional laboratory sessions. Between sessions they sometimes assign each other things to look up and things to think about. It might be that the lecturers in charge are too starry-eyed about their own initiative. Nonetheless, we have no desire to go back to the "catalogue of methods" approach to the teaching of analytical chemistry.

Chemistry educators are quick to claim the relevance of their discipline to the real world, but often take offence when it is pointed out that their teaching is not "real world" enough. Nevertheless, changes are always occurring in chemistry education, many of which answer in part the criticisms which are levelled at us. This is not a statement of complacency, rather an assertion that criticism can often lead, through an examination of what we do, to innovations and, hopefully, improvements.

References

1. Good examples are Reuben, B. G. and Burstall, M. L., *The Chemical Industry*, Longman, 1973; Witcoff, H. A. and Reuben, B. G., *Industrial Organic Chemicals in Perspective, Parts I and II*, Wiley, 1980.
2. Undergraduate Teaching Syllabus in Analytical Chemistry, Fourth Report of the

Committee of the Education and Training Group of the Analytical Division of the Royal Society of Chemistry, Anal. Proc., March 1982, pp. 104–109.
3. Pardue, H. L., Holistic Approach Urged for Teaching Analytical Chemistry, *Chemical and Engineering News*, Volume 61, 12 September 1983, pp. 37 38.

46
The Students' Project Programme

M. A. SETHU RAO

Karnataka State Council for Science and Technology, Bangalore, India

The Karnataka State Council for Science and Technology (KSCST) was established to assist the State in the application of Science and Technology to its development needs. In particular, the objectives of the State Council were defined as follows:

> To identify areas for the application of Science and Technology to the development needs, objectives and goals of Karnataka;
> To advise Government on the formulation of policies and measures which will promote such application;
> To promote effective co-ordination and to develop and foster communication and other links between centres of scientific and technological research, government agencies, farms, industries etc;
> To initiate, support and co-ordinate applied research programmes in universities and other institutions.

KSCST's objectives are achieved through the initiation, funding and implementation of projects. Throughout the three sequential phases of a project, viz, (1) problem identification, (2) generation of technological solutions, and (3) implementation and dissemination of solutions. This collaboration is achieved through the joint participation of the diffusers (Government policy makers and administrators) and generators (scientists, engineers and doctors) in the Working Groups set up for each project. This approach has the advantage of making the diffusers of technology aware of all the technical possibilities, while it simultaneously presents them with an opportunity to inform the generators of technology about policy constraints. The underlying philosophy is simple. Technology diffusion is primarily the responsibility of Government development agencies led by Government policy-makers and implementors; and technology generation is mainly achieved through scientists, doctors and engineers. Instead of establishing new institutions, KSCST seeks to harness the educational, scientific and technological infrastructure that already exists in the institutions of Karnataka.

Students' Project Programme (SPP)

At the start of its work in 1976, KSCST located its projects in the major institutions in the State such as the University of Agricultural Sciences, the Indian Institute of Science, the National Aeronautical Laboratory, and the Central Food Technological Research Institute. But to avoid concentration of the projects in these selected institutions in Bangalore and Mysore, and to involve a large number of institutions spread throughout the State, KSCST saw in the engineering colleges of Karnataka, an enormous reservoir of talent waiting to be harnessed in the interests of the State. KSCST decided to implement a programme — *the Students' Project Programme* (SPP) — to provide financial and academic support for projects undertaken by undergraduates in engineering colleges in Karnataka. Project work was already a compulsory curricula requirement for final year students but because of paucity of funds and ideas, most projects were either "on paper" or depended on software.

On implementing SPP, KSCST was confronted with a novel and unexpected situation. The students faced projects of a completely new type, sometimes requiring arduous visits to villages, discussions with rural artisans, acquiring basic data through painstaking measurements and laborious field surveys. The faculty guiding the projects were faced with problems for which they had no stock solutions. In other words, research had to be done.

SPP has a far-reaching significance. It links together the development efforts of the state with educational institutions, first the Engineering colleges and later the Medical and Fisheries colleges, so that relevant problems flow to the institutions. In generating "solutions" to such problems, the students and faculty are obliged to interact with the prospective beneficiaries of development, the people. This interaction leads to a two-way flow of benefits — the so-called educated get educated into the realities of our country and the underprivileged masses get their privileged elite to understand and, if possible, solve their problems. An attempt to apply Science and Technology to the problems of the people enriches Science and Technology.

The programme also has a major impact on the process of education. Confronted with real problems, the basic engineering and scientific knowledge becomes meaningful. Its power becomes highlighted and its limits become exposed. And this mechanism of education is cheaper, faster and more effective than merely trying to imitate the educational systems and practice of the advanced countries.

The environment around is the best teacher. But to be taught, there must be active involvement with this environment, and commitment to the task of improving it. Thus, SPP has turned out to be a major innovation in technical education.

Proposal generation and processing

Primarily, the projects to be carried out by students in Engineering, Fisheries and Medical colleges should promote KSCST's objectives. The criterion, therefore, is whether the project is relevant to the development needs of Karnataka. That is, does it lead to

(a) the satisfaction of basic human needs,
(b) self-reliance, and
(c) harmony with the environment.

The projects must be relevant to the basic needs of food, shelter, clothing, health, transport, and employment. They must lead to an increase of self-reliance (of villages, hoblies, blocks, taluks, districts and the state) through greater use and utilisation of local resources (materials, skills and technology). Problems for SPP project work may be identified by KSCST, the college faculty or students. KSCST regularly compiles lists of projects from the suggestions. The student projects finally selected are either new or alternative approaches/improvements to projects already executed or continuation of projects carried out in the same or other colleges during previous years.

Though the major criterion for choice of a project is the relevance of the project to the development of Karnataka, the selected project would, of course, have to be consistent with the academic requirements, with the interests of faculty and students, and with their equipment and time constraints. The following are some significant SPP projects.

1. Study of firewood and charcoal consumption of a village
2. Brick-making machine or soil-compacting machine
3. Low-cost tile roofing unit
4. Solar water-heater
5. Community biogas plant
6. Wood burning stove
7. Sisal decorticator
8. Rope-making machine
9. Low-cost saw mill
10. Coconut oil expeller
11. Seed sowing cum fertiliser driller
12. Performance study of centrifugal pump
13. Sample health survey of a village where there is no doctor
14. Evaluation of vaccination coverage in rural community
15. Studies on paddy-cum-fish culture
16. Growth response of carps to fishmeal substitutes enriched with sardine oil

Students can obtain guidance, including technical information, from KSCST personnel and from members of their own college. KSCST personnel visit students during the project, and funds are available (about US$200–500).

After evaluation, projects which show scope for further improvement, or are technically successful in themselves, are selected for exhibitions held in the participating colleges. These exhibitions are attended by industrialists, academics and entrepreneurs, and of course, the students. Of the 160 projects which were funded by KSCST in 1983–84, 30 were selected for seminars and 20 for exhibitions at the Bapuji Institute of Engineering and Technology at Davangere.

Many of the projects under SPP require further development, refining and field testing before they can be considered for commercialisation. In fact, some students become frustrated because their "work" is not immediately commercialised!

However these projects must meet the requirement of their course and only then can field testing, market research and product development be considered. In order to carry out these tasks, KSCST has initiated a programme of establishing Product Development Centres in some of the engineering colleges.

The first of these PDCs is now at the Shri Jayachamarajendra Engineering College at Mysore. The following are a few of the technologies ready for commercialisation at this PDC.

1. Electronic kit (a teaching aid)
2. Orthographic view projector
3. Surface grinding dynamometer and lathe tool dynamometer
4. (a) Power pack for linear and digital IC applications
    (b) Power pack for general applications
5. A deflossing machine
6. The design and fabrication of a half-wave rectifier

Problems and prospects

In the first year of the scheme 1977, 81 projects from 13 colleges were undertaken. By 1984, these numbers had doubled.

There are constraints on further progress: funding is becoming more difficult, so is monitoring the scheme for it is the concern of KSCST to maintain a high quality. It is hoped that other bodies will become involved in this work including the Khadi and Village Industries Board, the Agro-Industries Corporation and the Department of Non-Conventional Energy Sources. More Product Development Centres may be established to follow up on SPP work. Entrepreneurs must be made aware of SPP possibilities. The future has challenges. Do we accept them?

47

A Postgraduate Diploma Course in Industrial Research and Management

R. A. KULKARNI

C. C. Shroff Research Institute, Bombay, India

Is there a gap between the education provided by the university and the education needed for industry? A survey was conducted in 1975 and 70 chemical and allied industries responded. Based on this, attempts were made to change the curriculum of the university B.Sc. course in chemistry, but without success. It was therefore decided to start a postgraduate diploma course to bridge the gap.

The curriculum was planned jointly with academic scientists, technologists and managers from industry, and it is revised every year from the experience gained. The faculty comes from industry and from the university, and frequent exchanges of the faculty are organised. The course is spread over one year. It is part-time, but is equivalent to about two months full-time. The students come with different experiences: some straight from university, others from a short while in industry, others have served many years in industry and use the course as an in-service refreshment.

The work contains courses in chemical engineering (the scale-up operation), quality control techniques (analytical chemistry), principles of management and economic factors, the latter being very important for people from small industries where failure is more often due to lack of financial rather than technological acumen. Much of the teaching is done as project and problem-solving work and the students work on real-life problems from industry. The practical component is also important, giving "hands-on" training on sophisticated expensive instruments as well as pilot plants for scaling-up operations, techniques which students have not had an opportunity of using earlier.

So far over 300 students have been awarded the diploma: they are employed immediately and companies clearly prefer diploma holders to other graduates, even ones with better academic qualifications. The success of the course can be gauged from the requests received from industrial companies for the Institute to let them know of new diploma holders.

48

Development of Resources for Science Teaching: an Approach for Encouraging Self Reliance

K. V. SANE

University of Delhi, India

The rising prices of equipment and chemicals are a matter of concern to all science educators because of the difficulties they are creating in the organisation of laboratory teaching.

One way of tackling this situation — which is becoming desperate in developing countries — is to treat equipment and chemicals as a part of resource material which should be developed by teachers and students. This idea can be made feasible by introducing Project Work, with an industrial and a technological bias, in the existing science curricula. As an illustration, this paper describes the use of Student Projects in the area of micro-electronics for development of low-cost equipment for teaching chemistry.

Delhi Project

A pilot project for design and fabrication of low cost equipment for chemical education was initiated at the University of Delhi in late 1979 by the Committee on Teaching of Chemistry (CTC) of the International Union of Pure and Applied Chemistry (IUPAC) in collaboration with Unesco and the Committee on Teaching of Science (CTS) of the International Council of Scientific Unions (ICSU). The general and the specific as well as the technical and non-technical aspects of the Project have been reported elsewhere.[1-6] It will therefore suffice to first outline the essence of the project and then proceed to examine such aspects which have relevance to some of the wider problems in science education.

An interdisciplinary team consisting of chemists, physicists and technicians — teachers as well as students — has evolved at the University of Delhi in the last 5 years to give the Project a firm direction. The team has so far designed low-cost versions of several instruments, accessories and experiments suitable for use at the junior university/college and senior school level. An essential feature has been the involvement of a large number of students through informal project work during vacations and holidays. Another feature has been the exclusive use of made-in-India components. Since none of the workers — teachers and students — have any prior training in electronics and instrumentation everyone is learning-by-doing which has helped to create an atmosphere where the stress is on simplicity.

Development work

The instruments chosen for development include pH meter, conducto-meter, colorimeter, polarimeter, thermometer and polarograph; the accessories include electrodes, conductance cells, thermocole-based thermos flasks, and magnetic stirrers; and experiments include selections from different branches like organic, inorganic, physical, biological, industrial, agricultural and environmental chemistry. Within a given piece of apparatus, several designs have been made to suit different requirements. For example, eight models of a pH meter are available which accommodate different options like battery/mains operation, low/high impedance, high/medium accuracy and low/ultra low cost. Similarly a conductance cell has designs which can be made with/without soldering facility. The prices range between US$8 for a very simple conductometer and polarimeter to US$50 for an all purpose high-accuracy pH meter and a good performance colorimeter. Almost all the accessories, with exception of a magnetic stirrer, are virtually zero cost since they are made from discarded materials like carbon rods of discharge dry cells, thermocole packings etc. etc.

Propagation work

The development work described above is one face of the Project; the other face is the propagation work. As long as the activity is limited to a single group and a few instruments, there is little doubt that it will have a negligible impact. The crux of good teaching and learning is attitude and not instruments, chemicals or similar materials. Attitudes however cannot be cultivated by mere platitudes; it is necessary to provide a framework where an enthusiastic, intelligent teacher and/or student has an opportunity to experiment, to create and to evaluate. The use of locally-available inexpensive, durable, safe and versatile electronic materials offer today unparalleled opportunities to initiate similar projects everywhere in the Third World countries. To catalyse this process, a series of Workshops have been held where participants are introduced to the philosophy of the Project followed by laboratory sessions involving designing, fabrication and testing of simple instruments. Beginning with an International Workshop in Madras in April, 1981, five such Workshops have been held in India with the support of national agencies and an equal number have been held outside (namely Brazil, Guyana, Denmark, France, Bangladesh) with the support of international and regional agencies.

It is important to emphasise that the Project — either in its developmental aspect or in its propagation aspect — is not attempting anything new. Chemical literature is replete with references where electronic devices like an operational amplifier and electrodic materials like a carbon electrode are employed for teaching and research purposes. If the Project can claim any novelty at all, it is in the sum total of its efforts and not in the individual parts. To appreciate this point, it is useful to examine the key features of the Project.

Key features

(1) International sponsorship: The sponsorship by professional organisations like the IUPAC and ICSU has helped to highlight the urgency of the problem, provided avenues for global exchange of information and international collaboration and offered expert scrutiny. There is no doubt that international organisations can be a very effective catalyst for initiating action in many problem areas in science education of relevance to developing countries, if not to the whole world.

(2) Interdisciplinary core group: A joint effort by chemists, physicists and technicians enables a balanced consideration of theoretical and practical matters and of fundamental and applied aspects. It also enables development of a package where the various pieces match with each other in cost, convenience and performance. To illustrate this point, consider the example of a pH meter. It is easy for a physicist to develop a circuit, for a chemist to make an electrode system, and for a technician to design the layout. However, unless the three parts are compatible in pricing accuracy and convenience, the overall set-up has not achieved an optimum design. (Of course, there are some versatile individuals who are a chemist, a physicist and an engineer rolled into one but since samples of such exotic species are rare, ordinary mortals have to learn to combine with each other.)

(3) Emphasis on quality: The term low-cost has unfortunately become synonymous with low quality. It is extremely important that in a Project like this quality should never be compromised. The objective is not to produce an apparatus which is merely cheap but to make one which in addition meets the demands of accuracy and precision required at a particular level (e.g. school, college) and for a particular purpose (e.g. titration, absolute measurement).

(4) Participation of students and teachers: This is one of the crucial features. The use of teachers and students in designing, fabrication, testing and evaluation creates a sense of involvement besides generating confidence and expertise required for more ambitious efforts. The ability to maintain, repair and modify routine equipment is an important offshoot.

(5) Linking training with production: In a majority of Workshops the participants have not only been introduced to the circuit design, its operation and functioning, but they have been encouraged to assemble at least one complete unit. Thus in the week-long Hyderabad (November '82) and Guyana (August '83) Workshops, participating teachers made about twenty instruments while in the three-week Delhi (June '83) Workshop, 12 students made 50 instruments. The fabricated equipment has been distributed to appropriate institutions under the auspices of the Workshop-sponsoring agency.

(6) Reliance on locally-available materials: The long-term and an assured growth of the Project demands that it depends only on easily available local materials. From an academic viewpoint also, this is a useful restriction because it teaches how to attack a problem subject to various constraints.

Many developing countries (e.g. India) are now witnessing the growth of electronics and materials technology which heralded the space age in the developed countries in the sixties. The seeds of these technologies, namely the electronic components and a variety of new materials, are now either made or are easily available in most of the developing countries. These components and materials are (generally speaking) cheaper, more versatile rugged and easier to use than their earlier counterparts. Some examples are:

(i)   the ICs (Integrated Circuits) like the operational amplifiers (Opamps) which can be used to construct simple circuits at a reasonably low cost;

(ii) transducers like thermistors, phototransistors etc. which can convert variables like temperature and light intensity to electrical quantities (e.g. resistance, current) thus extending the usability of electrical variables;

(iii) materials like polyethylene, polyvinyl chloride which can be used in a variety of ways since they can be easily manipulated;

(iv) discarded items like discharged dry cells, and thermocole packings which provide useful products like electrode material and insulation material respectively.

It must be pointed out that this is only the beginning. Components and products far more versatile, powerful and economical than these are already commonplace in the developed countries. With the appreciable decrease that has taken place in the timelag for the transfer of technology, these devices will be available in developing countries sooner and not later. It can thus be concluded that conditions are ripe for academics in the developing countries to seize the chance of harnessing technology to make education in science comprehensive, exciting and purposeful. However, if these happenings are to make a substantial impact on the educational scene, a strategy has to be devised which views instrumentation not as an end in itself but as a means of catalysing certain much needed changes in the teaching methodology.

In a nutshell, the most significant exercise about the work appears to be its attempt at various types of interfacing (e.g. international/national; low-cost/high-quality; chemistry/physics; teachers/students; training/production) *simultaneously*. Each of these interfaces is felt to be sufficiently crucial that neglect of any of them will distort the underlying philosophy and manner of operation in a significant way.

Future

As long as the locally-produced equipment Project described above remains confined to a single group (e.g. the Delhi group) a single subject (i.e. chemistry) and a few instruments (i.e. a pH meter, a conductometer), the activity will not have a large-scale impact. It is necessary therefore to initiate new groups, to spread the approach to sister sciences, and to systematically build as much equipment as is possible for teaching of science at all levels. It also needs to be investigated whether the six key features listed earlier can provide a base for parallel projects to generate the resources needed for science education today like audio-visual material, software for microcomputers, low-cost chemicals etc. etc. In other words, it is necessary to take gradual steps to convert the Project into a Movement by involving an increasing number of individuals, groups and agencies all over the world.

There is one point pertaining to the future which requires careful thought. This concerns the mode of support for such a Project. One possibility is that work like this relies on grants from national/international agencies. Although this is a standard practice followed everywhere for supporting academic programmes in educational institutions, the disadvantage is that it tends to make an activity parasitic. Another possibility is to set up a small-scale industry whose objective is to cater to the specific needs of science education by making items like equipment, chemicals, audio-visual, software etc. etc. This is a very attractive idea particularly in a big market country like India as it seems to promise financial self-sufficiency to projects aiming to achieve self-reliance in generating resources (e.g. equipment etc.) for science education. However, there is a real danger that the profit motive may overtake the academic objectives. The involvement of teachers/students may start decreasing and even disappear; the prices may start moving up and even equal the commercial models. If and when this happens the exercise will become self-defeating. The third possibility, based on the principle of golden mean, envisages that development and propagation will be the responsibility of an academic wing but production and distribution of material (i.e. equipment, chemical, software etc.) will be entrusted to a semi-commercial wing. The former operating on a non-profit basis will be supported in the usual manner by national/international funding agencies. Its objective is to motivate teachers, train students and provide a format for small-scale innovations and problem-solving. The semi-commercial wing will employ skilled and unskilled workers on a full-time basis and it will also utilise a large number of students on a casual basis during holidays and vacations. The students will thus get an opportunity to undergo practical training under realistic factory conditions. It is expected that this wing will supply items to educational institutions at a nominal profit which will be used for wages of full-time and part-time workers and for stipends of the students.

The model favoured has been attempted, with success, on a limited scale by the Delhi Group. Over 200 instruments have been supplied to educational institutions in India and abroad using students working in vacations, under one technical supervisor. The virtue of the model is that it attempts to get the best of the two worlds — academic and industrial. Of course many problems are expected to arise when the idea is formalised. This danger notwithstanding, it is felt that the model is worth pursuing in collaboration with a suitable Industrial House because it offers a bridge — albeit a narrow and restricted one — between the academic practice and the industrial practice and between the prototype effort and the production set-up. If the attempt is even partially successful, it can provide a viable first step towards vitalisation of science teaching.

Summary

Two of the more important problems confronting teaching of science in developing countries are (i) lack of resource material and (ii) lack of a framework where student/teachers can develop problem-solving skills and attempt simple innovations. It has been found that a beginning towards a solution to the twin problems can be made by initiating projects where students/teachers undertake developmental work using local materials. However, if such a beginning is to have a substantial impact it is necessary to organise several groups working in a co-ordinated manner. It is also necessary to establish a link with the industry so that large-scale production of successful items can be undertaken for distribution/sale to educational institutions.

References

1. Bhattacharjee, R., Sane, K., Sane, K. V. and Srivastava, P. K. *Indian Journal of Chemical Education*, **8**; 1 (1981).
2. *Proceedings of the 6th International Conference on Chemical Education*, Maryland, U.S.A., 1981 (Contains papers by K. Sane, K. V. Sane and P. K. Srivastava).
3. Waddington, D. J. *International Newsletter on Chemical Education*, **17**, 6 (1982).
4. Sane, K. V. *International Newsletter on Chemical Education*, **17**, 9 (1982).
5. *Locally Produced and Low Cost Equipment and Experiments for Chemistry Teaching*, Volume 1, IUPAC/Unesco Publication (1982).
5. *Proceedings of the Workshop on Low Cost Equipment*, Royal Danish School of Education, Copenhagen (1983). (Papers by K. V. Sane, C. K. Seth and P. K. Srivastava).

Note

Two poster papers were presented at the Bangalore Conference. One by R. K. Bali (D-157 Vivek Vihar, Delhi 110 032, India) described a student pH meter and a conductivity meter for use in schools. Details can be obtained from R. K. Bali. The other described a low-cost double-beam spectrophotometer and was presented by F. Abildgard, J. J. Christiansen, C. K. Seth and E. W. Thulstrup. Details can be obtained from E. W. Thulstrup, Royal Danish School of Educational Studies, Emdrupvej 115B, DK-2400 Copenhagen NV, Denmark.

49

In Praise of Projects

J. A. LEISTEN

Chancellor College, University of Malawi

Marking a recent test the writer found that several of his students had calculated the frequency of a spectral line from a single molecular energy state; and examination scripts from another university contained some otherwise reasonable answers to a question on the dissociation of chlorine which began with the "equation": $Cl_2 \rightarrow 2Cl^-$. Markers of present-day chemistry examinations are quite accustomed to mistakes of this kind, which suggest a fragmentary and not very genuine understanding of the subject. Our students seldom seem to emerge from lecture courses with the coherent kind of knowledge needed for successful practice. Yet we know that in a few years time many of these insecure students will have changed into competent chemists, and this leads naturally to two questions: what causes the change, and is there any way of teaching that might hurry it along?

Obviously, chemists learn by experience. They are able to avoid mistakes and overcome obstacles as a result of many previous struggles of trial and error. The writer believes that project work comes closer to providing such experiences in schools and universities than any of the other learning methods. If methyl orange is used for titrating acetic acid with alkali in the course of a project, the mistake must be identified and corrected before progress can be resumed and the objective achieved. The same mistake in an examination merely loses a mark or two! In a project the student is on his own. In laboratory courses he may do as the others do, or as the instruction sheet tells him. Of course all learning methods have their particular functions, but none of them can rival projects for encouraging students to face realities, and so develop a working, as opposed to a bookish, approach to chemistry.

Many teachers might agree with this assessment, and yet feel that projects have only marginal importance in teaching chemistry because they are too demanding on staff and laboratory resources. The present paper is concerned with ways of overcoming these constraints so that a broader range of students can benefit from project work. Most of the ideas have been tried at Chancellor College.

Long projects

It is quite usual for students in the final year of an honours degree to have a full year project, and this is the case at Chancellor College. Recent projects that have been devised and supervised by colleagues include the solubilisation of potassium in feldspar as a potential source of fertiliser,[1] the properties of Malawian tea-seed oil with a view to its utilisation,[2] and a regional survey of trace iodides in drinking water.[3] (In the latter project there was an interesting analytical method depending on catalysis of a chemical reaction by the iodide ion, and the results were particularly needed by the public health authorities.) It is easy to imagine the stimulus received by a student who successfully manages a practical application of chemistry to his country's needs as in these examples.

Problems arise however when a student finds his project too difficult. The supervisor, not wishing to be hard on someone who has reached the final year of a lengthy education, may then shoulder too great a share of the burden, and the project may become educationally less effective and perhaps even discouraging. This danger can be reduced by introducing students to project work more gently, and in particular by setting shorter, less exacting, more closely supervised projects in the penultimate year of the degree course.

Economical projects

With our final (fifth) year students already taking up resources it is difficult to provide projects for the greater numbers in the penultimate (fourth) year, and so a specially economical scheme was created for them. It has already been described,[4] and what follows is a summary, with some further suggestions.

The central idea is to have each student working on a different aspect of the same theme. Our theme is a chemical reaction, the iodination of acetone. The administration of the laboratory is immediately simplified because many of the solutions and chemicals are common to all the projects; but the most important advantage of the scheme lies in the supervision. Students carry out the projects in six weekly four-hour periods, and apart from the first one, when a colleague's help is required, a single lecturer who is well acquainted with the subject can supervise up to twenty students by himself. Different students may use different experimental methods (spectrophotometry, conductivity, gas-liquid chromatography, calorimetry, potentiometry) and study different aspects of the reaction (catalysis, autocatalysis, the effect of temperature, quenching methods, and so on). At the end, each student gives a talk on his or her work in a one-day symposium. The common theme makes it easier for students to communicate their results to each other, and as a further help the supervisor sees that the talks follow in a logical order.

It perhaps illustrates a trend in chemical education that if the theme for these economical, intermediate, projects had been chosen today instead of 15 years ago it would have been in applied chemistry. Today's theme might be a local industry, such as a sugar refinery. One project might be on the estimation of mineral salts in sugar solutions, and another on the removal of such salts by ion exchange resins. Ion exchange of cations lowers pH, and so these two projects might be followed at the symposium by one on the effect of pH on the inversion of sucrose, and its possible relevance in sugar refining. It would be difficult at first to find suitable projects for twenty students or so, but experience with the iodination of acetone shows that almost every laboratory period generates a new idea. A particular advantage of choosing an applied theme would be the extra link with industry, and it would not be long before some of the students were working on projects of real interest to the chemists in the factory.

Laboratory problems

If project work is important in the later years of a chemistry course it must surely have some value earlier on, and the process of introducing project work in progressively increasing steps could well begin when students start chemistry. These thoughts led to the development of a laboratory problems course for use in the second and third years at Chancellor College. In the following description there are two intentions: the first, to enable teachers to assess the course; and the second, to pass on some results of our experience to any who may wish to set up a similar course elsewhere. Both intentions can be served by following in imagination the progress of a typical class.

It will be seen from this description that close contact between class and supervisor is essential, and we therefore divide the second year class into groups of ten. Imagine now that one such group enters the laboratory for its fortnightly problem. Each student finds two unknowns, a set of test materials, and a laboratory problem (item 98) typed at the top of a report sheet.

ITEM 98. Think of as many ways as possible of distinguishing between compounds with the structures $CH_3(CH_2)_{16}CO_2H$ and $HO_2C(CH_2)_2$ CO_2H using only the materials provided on the bench. Then identify A and B, which are samples of the two compounds, using all the different methods.

The supervisor allows 45 minutes for the devising of methods, students being allowed to consult notes and books but not each other. After 15 minutes perhaps the first student shows the supervisor the three methods he has thought of, and there begins a written dialogue, on the student's report sheet, which lasts until the end of the two-hour problem class.

The first student has decided that a match-head amount of succinic acid will need more drops of decimolar sodium hydroxide to neutralise it than a similar quantity of stearic acid. Stearic acid should burn with a smokier flame. It should be less soluble in water. The supervisor writes *Good. Can you think of more methods?* After a while the student realises that succinic acid should have a higher melting point on account of hydrogen bonding. The supervisor writes *Yes, but how will you compare the melting points?* (Students find different answers to this question even though the only apparatus available is a spoon spatula and a burner.) The student now needs some hints, so the supervisor writes *Solubility in non-polar solvents?* One of the test reagents is bromine in carbon tetrachloride, and some students realise that this will serve as a non-polar solvent, and that stearic acid will dissolve in it more readily than succinic acid. Perhaps one further hint will lead to a sixth way of distinguishing the two samples, the production of soap bubbles when a neutralised solution of stearic acid is shaken in the test tube.

The supervisor now lists all six methods on the board so that everyone has the same chance in the practical work. Students at this level are often more sceptical about chemistry than their teachers assume, and it is a useful feature of item 98 that six independent methods are found to lead, eventually, to one and the same conclusion. As in the first phase of the problem, so in the practical work, some students will be more effective than others, some will report their results more clearly, and some will be quicker. Those who finish early enough are given a supplementary question: *Think of other ways of distinguishing these two acids, using additional reagents or equipment.* Universal indicator would reveal succinic acid as the stronger acid through the inductive effect of neighbouring carboxyl, infrared spectroscopy would show a large difference in the proportions of CO_2-H to C-H, potentiometric titration would give two end points for succinic acid, and so on.

At the end of the two-hour period the report sheets are handed in for marking, and they are not returned because it is necessary to safeguard the problem bank. Students leave laboratory problem classes with nothing tangible in the form of laboratory or lecture notes. Yet even over a short period of weeks there is a feeling of gained confidence that other types of teaching can hardly match.

The example described give a fair impression of the laboratory problem course when it is properly established, but it does not bring out certain pitfalls and difficulties that are encountered in the early stages. When the course was first introduced, to students who were used to tests of their knowledge but not used to tests of their performance, the problems were resented. They seemed to have no purpose but to show up inadequacies. The fact that the course is now accepted, and even welcomed, is undoubtedly because the prevailing attitude is not criticism of the students but self-criticism by the item designers and supervisors, and this is perhaps the one general requirement for mounting a course of this kind. A few examples of how it operates might be helpful.

If a new problem defeats all the students it is the designer who is to blame, not the class. Enthusiasm may impair the problem setter's objectivity, but after a time the fault in an unsuccessful item is usually clear. One early item virtually required students to re-invent the method of back titration, an exercise that is clearly too demanding. If the basic idea is good, re-structuring or even just re-wording can usually save an unsuccessful problem. Alternatively we may move it to the third year course, in which the problems are more difficult and/or based on a greater knowledge of chemistry.

Laboratory problems are an exercise in the application rather than the acquisition of knowledge. The bank has sections on different topics, and there are for example problems on stoichiometry, conductance, solubility, oxidation-reduction, and light absorption; but all of them are based carefully upon facts with which the students are familiar. The organic problems are based largely on just seven chemical tests, so students are advised to study these tests and understand the theory of them before they are given any organic items. It cannot be assumed that students are familiar with everything in their lecture notes, and in developing viable new problems we find it impossible to avoid trial and error.

A good laboratory problem has sufficient flexibility to accommodate the ability range of the class, and this is well illustrated by item 98. A weak student will perhaps think of two methods for distinguishing the solids, and will get more with sufficiently broad hints. A student who thinks of four methods without help and then finishes the practical work in time to suggest one or two additional methods might qualify for full marks. A problem which totally defeats some of the students and occupies the quickest ones for only half the period is good for neither group; it must go back to the drawing board.

The most difficult aspect of laboratory problems is the supervision. The supervisor must try to give just the right help, when it is really needed, and to do this he must assess the student's suggestions accurately, sympathetically, and also quickly since there are ten students to converse with. It is difficult because students often produce the unexpected idea, and more often still they report the germ of an idea in a way which makes it hard to recognise. (Laboratory problems provide a good opportunity for improving written expression.) Success depends upon encouragement, and *Good* is the supervisor's most used word. It is active teaching which often strains the patience, but this is rewarded as the course goes on and the supervisor realises that students are tackling problems which in the early weeks would have left them just staring at the bench.

Conclusion

There must be many other ways of achieving the same objective, but our economical projects and laboratory problems have helped to show that a small university department can, without any extra resources, carry on systematic, step-by-step education of the project type extending throughout the students' undergraduate years.

It might be fitting to conclude with the words of a technical manager from Blantyre, Malawi, who had come to the University to recruit new graduates: "We want a chemist who can not only look after the routine tests, but can also sort out the problems that arise." How many prospective employers around the world would echo these words? And can anyone doubt the value of projects in educating this desirable type of graduate?

References

1. Dolozi, M. B., Goddard, D. R. and Masamba, W. R. L. *Luso: J. Sci. Tech. (Malaŵi)*, **4**(2), in the press.
2. Ridley, R. G., Mjojo, C. C., Likaku, C., Kambuwa, L. and Cloughley, J. B. *ibid.*, **4**(1), 39, 1983.
3. Guta, C. W., and Rogers, N. E. *ibid.*, **3**(2), 89, 1982.
4. Chokotho, N. C. and Leisten, J. A. *J. Chem. Educat.*, **58**, 490, 1981.

50

Teaching Organic and Polymer Chemistry in an Industrial Context

H. A. WITTCOFF

Chem. Systems International Ltd., London, UK

In both developed and undeveloped countries there is a gap between industrial and academic chemistry. This problem, serious in developed countries, can become critical in Third World nations. It is true that a chemical industry in a developing nation can be instituted by way of purchased technology and turn-key plants. But if these turn-key plants are operated mechanically, without understanding, and if the management of the companies that operate these plants does not understand the chemistry involved and related chemistry that is developing in the world around them, performance will at best be pedestrian and at worst completely inefficient.

The methyl isocyanate disaster — some might call it Armageddon — in Bhopal in India, emphasises this point. It is essential that chemists and engineers in developing countries be trained so that they can demand efficient operation. That lack of efficiency can be life-threatening is emphasised manyfold by this tragedy.

All my experience of organising courses in industrial chemistry indicates that it is important that chemists have an understanding of the implications and constraints of industry *before* they start working in industry. They would be even more effective if they understood what particular niche they might occupy in the vast expanse that comprises industrial chemistry On-the-job training is, of course, important; but it is far more effective if the chemist or engineer has studied the discipline of industrial chemistry before he comes to the job.

This gap between industry and academia has motivated the development of a course in which the actual chemistry basic to industrial practice is taught.

It is amazing, indeed, that this chemistry, most of which today has theoretical underpinnings, is not found in the widely used organic chemistry texts. The course is based on several understandings, one of which is that 90% of all of the organic chemicals produced are derived from seven basic raw materials: ethylene, propylene, the C_4 unsaturates, benzene, toluene, xylenes, and methane. It is important, accordingly, to find out where in the petroleum refinery the seven basic raw materials come from and thereafter to learn the chemistry associated with each of these. By using this approach it is possible to treat industrial organic chemistry as a discipline.

Three refinery reactions — thermal cracking, steam cracking, and catalytic reforming — produce the basic chemicals. These processes are covered in detail. Ethylene is the most important of the chemicals. Roughly 40% of all petrochemicals are produced from it. Even so there are no more than 15 chemical reactions that need be learned in order to cover the bulk of ethylene chemistry. These include polymerisation to polyethylenes and reactions such as oxychlorination to make vinyl chloride, the Wacker reactions to make acetaldehyde and vinyl acetate, and the reactions which lead to styrene, ethylene oxide, ethylene glycol, and ethylene oligomers. A study of the polymers of ethylene provides an opportunity to discuss free radical and Ziegler polymerisation as well as to introduce many polymer concepts such as crystallinity, copolymerisation, source of strength in polymers, and polymer properties generally.

Oxychlorination provides insight into how HCl can be converted into chlorine *in situ*. The Wacker reaction demonstrate how a stoichiometric reactant, palladium, can be converted into a catalyst. Each reaction studied demonstrates a key point of this sort.

This same approach is followed with each of the seven basic raw materials. Propylene chemistry points up the important fact that a chemical's low cost can motivate highly creative chemical reactions. Oxidation of propylene to acrylic acid and ammoxidation to acrylonitrile are cases in point. The "two-for-one" reactions based on propylene which produce phenol and acetone or propylene oxide and styrene are further examples. Metathesis, one of the most creative reactions in organic chemistry, was first done with propylene. Although this was never commercialised, the metathesis reaction is a key step in another ingenious process in which ethylene is oligomerised to give only C_{11} and C_{15} chain lengths, the most desirable for surfactants.

C_4 chemistry stresses elastomers since butadiene is a major component of synthetic rubbers. It is also the starting material for an important adiponitrile synthesis by intriguing chemistry: the addition of two mols of HCN stepwise to butadiene.

Benzene chemistry relates to the production of styrene, phenol and two nylon components, adipic acid and caprolactam. It also leads to a discussion of herbicides and dioxin.

Toluene chemistry points up the important fact that its excess availability has not motivated the same successful creative chemistry as has propylene's low cost. Its major use is for conversion to benzene by hydrodealkylation.

It is the basis for toluene diisocyanate which gives an opportunity for the discussion of urethane chemistry.

Xylene chemistry centres about the dibasic acids resulting from xylenes' oxidation. Poly(ethylene terephthalate) and modern engineering polymers are related to xylene chemistry.

Methane chemistry leads to the formerly important chemistry of acetylene. Most important it is the basis for C_1 chemistry, the reactions of CO and H_2 or synthesis gas, which may someday replace much of the chemistry we practise today. The synthesis gas can, of course, come from coal as well as from methane or other petroleum sources. From it we can make aliphatic and aromatic hydrocarbons. These will be chemical feeds when petroleum is exhausted, although the switch-over will be tortuous and costly.

C_1 chemistry is practised today with petroleum-based synthesis gas for the preparation of ammonia and methyl alcohol. More recently chemists have learned to make acetic acid from methyl alcohol and CO as well as acetic anhydride from methyl acetate and CO.

C_1 chemistry requires an understanding of homogeneous catalysis. Other catalyst advances — metalloorganic, hybrid, zeolites, dual function, cluster, solid acid, enzymes, phase transfer, and catalysts for asymmetric synthesis — are described.

Application of chemicals not only in polymers but also in other areas such as surfactants and pharmaceuticals is stressed. Finally, the course covers the chemicals based on non-petroleum sources: coal, carbohydrates, and fats and oils. The chemistry practised with these materials is unique and will become more important as the availability of petroleum declines.

The course is punctuated by historical perspective, anecdotes and humanism, and attempts to teach science as liberal arts. It has been given in India, Brazil, Nigeria, Israel, China, France, UK, Puerto Rico, Thailand and the United States, and has been well received in both developed and developing countries.

51

Workshops on Industrial Organic Chemistry in India

R. A. KULKARNI

C. C. Shroff Research Institute, Bombay, India

H. A. WITTCOFF

Chem. Systems International Ltd., London, UK

India is a developing country, but unlike many developing countries it already has a sophisticated chemical industry as demonstrated by the fact that there are 3 crackers for ethylene production on the one hand and a highly developed pesticide industry on the other, showing that the industry encompasses everything from basic heavy chemicals to performance chemicals. That the Indian chemical industry is supported by its educational system is indicated by the fact that there are an estimated 2.0 million (1980) active Science & Technology personnel in India today and PhDs were granted at the rate of 2862 for Science subjects and 186 for Engineering & Technology in the year 1981–82. Some 60–70% of these belong to the discipline of chemistry. At the same time, 567,000 students in science for bachelor's and master's degrees are enrolled in the Universities, out of which almost 50% are awarded degrees, again of which 60–70% are chemists.

To maintain and expand this promising industry, it is necessary that chemists and engineers have a good understanding of how the chemical industry operates in the developed countries and what advances are being made. There are sources for this information in the chemical journals and books. But for the most part, these do not give a broad overview of how the industry operates and do not indicate to a chemist in India where his particular interests fit in the overall industrial chemical spectrum.

In developed countries, an infrastructure exists which the chemists learn about and use once they enter the industry. This infrastructure encompasses a vast gamut from the philosophy motivating research to the efficient operation of a stockroom. This infrastructure is rudimentary in less developed countries and must be built up by the young chemists and engineers entering the industry. Thus, the needs for industrial chemists and engineers in all countries can be summarised as follows:

1. To understand, as a discipline, basic industrial chemistry. This is not taught in the schools in developing countries to any greater extent than it is in the developed countries.
2. To understand how industrial organic chemistry is practised in the developed nations.
3. To understand what advances are being made in industrial chemistry and to understand how these advances may affect chemical industry in developing countries in the future.
4. To understand the problems facing the industry and to gain some feel for how these problems may be solved.

With these needs in mind, three workshops on industrial organic chemistry have been organised by the authors in Bombay and Baroda during 1982–84.

The course content followed the pattern described in Chapter 00. Every effort was made to illustrate the similarities and differences in the United States and Europe and the developing countries. For example, acetaldehyde is made in India, Brazil and China by an older process, the oxidation of ethanol. In the UK, ethylene hydration by the Wacker reaction is practised. But from the point of view of both economics and the "benefit-to-nation" concept, the older process is better for India and Brazil with their indigenous molasses or sugar cane juice for fermentation to alcohol. When ethylene made from indigenous petroleum becomes plentiful and cheap, it may be advantageous for India and Brazil to use the Wacker process.

Ethylene from fermentation alcohol is another example. Western Europe and Japan would not entertain such a process. Nor would Saudi Arabia with its plentiful supply of natural gas. But for countries rich in sugar cane, the older process can prove economical if it spares hard currency by reducing imports of oil, especially if the ethylene is to be used within the country.

Nowhere is the "benefit to nations" concept better demonstrated than in New Zealand where a process will soon be in operation for converting methanol from cheap natural gas to gasoline. The process would be far from economic in the United States. Not so in New Zealand where a cheap, plentiful resource, natural gas, is used to decrease imports of petroleum and thus to provide the country with more favourable balance of payments.

Some developing countries are rich in plants which provide specialised chemicals for the rest of the world. Although these are produced in small amounts, they are important to their countries' economies. For example, the carbohydrate polymer, guar, is indigenous to India. Both in India and Mexico, the barbasco root, which yields diosgenin for conversion to cortisone, is found. Brazil is the world's only source of pilocarpine. An attempt, accordingly, was made to stress these contributions from the chemistry of developing countries. Since the course was being presented in India, emphasis was placed on how the Indian chemical industry compares with that of developed nations and what challenges lay ahead in order to make the Indian industry more sophisticated and more productive. Emphasis was also placed on the fact that there are usually several processes available to manufacture a given chemical. The best process is not necessarily the latest but, as indicated above, one which can best be adapted to local conditions.

Since the chemical industry is considered "mature", there are certain problems associated with it. What these problems are were explained in the course and the point was stressed that a problem in a developed nation can provide opportunities for developing nations. The methyl isocyanate disaster in Bhopal emphasises how important is the inclusion of topics in the course illustrating the social and ethical obligations of the industry.

The importance of government regulations, the need for more accurate determination of degree of toxicity, toxic waste disposal, and clear air and water regulations were discussed. Examples were cited of the difficulty of judging toxicity (DDT, Dioxin, Saccharin, Cyclamate, etc.) Overall, the importance of making judgements of risk-versus-benefit was stressed, judgements which are not necessarily universal for all countries. Future courses will describe two equally profitable methods for making carbaryl, the product manufactured at Bhopal, one of which does not make use of methyl isocyanate. Comparative advantages of the two processes will be probed from the points of view of capital investment, safety and profitability.

Although the courses deal largely with chemistry, the economics of production are emphasised. Volumes of chemicals produced and trends in production of specific materials are described. The students learn how the cost and selling price of a chemical is computed and what contribution this makes to choice of a process. As an example, there are three processes for the production of hexamethylene diamine, the most expensive starting with adipic acid. The intermediate cost process starts with acrylonitrile and the cheapest with butadiene and HCN. Presumably each process is economic under the conditions where it is used. These complex circumstances are probed.

There are several new processes for styrene manufacture, all of which involve the interaction of methanol with toluene. Why this reaction is not used in the United States, but why it might be useful in — for example — China, provides interesting pedagogy.

While English is a language basic to both India and the United States, differences in usage and pronunciation can create communication problems. This is a point worthy of serious consideration wherever in the world teaching is done. To alleviate as much of the communication problem as possible, teaching was done completely with transparencies. All were previously reproduced and a set of each was given to each participant. In this way, key words could be perceived both orally and visually. Note-taking was minimised. Key points were repeated several times, and every attempt was made to convert questions into pedagogical experiences.

Future workshops will concentrate on other themes — pharmaceuticals; agrochemicals; natural products.

Suggested follow-through

Workshops play an important role everywhere. However, because of cost and lack of trained teachers, only a limited number can be sponsored each year. If some of the important concepts of industrial chemistry could be incorporated into university curricula, the graduate would benefit.

In the United States, a pattern has been set which may be useful in other countries. Through a National Science Foundation-funded and American Association for the Advancement of Science-operated programme called Chautauqua, over 500 college and university teachers were given a three-day (22 hour) workshop on industrial organic chemistry. Although 500 is a small proportion of the total, it is safe to say that most of these people will include industrial chemical information into their organic courses. Such workshops could be organised to train university teachers in developing countries who can, in turn, transmit this to students.

A concurrent proposal is to assemble small groups (10–25 persons) of industrial chemists who have an inclination and capability to teach. These persons will be trained in a five-day workshop to teach industrial chemistry to others, primarily to their colleagues. Teaching materials would be made available and reference material described. This pyramid effect can multiply the effectiveness of the number of people who are available to impart knowledge relating to industrial chemistry.

52
Some Concluding Comments

K. KING
University of Edinburgh, Edinburgh EH8 9JT, UK
D. McCORMICK
Manchester Polytechnic, Manchester M1 5GD, UK

From Dr K. King

It is important to be aware of the crucially significant distinction between the world of education, the world of training and the world of production. The process of trying to introduce the "feel" or "relevance" of one sphere into another (for example, putting something of the world of production into schools and colleges) usually means that the element introduced is powerfully affected by the wider environment of education. It follows that "production in schools" has little similarity with production in industry or agriculture.

Achi (Chapter 42), Sethu Rao (Chapter 46) and Kulkarni (Chapter 47) all take as a starting point the problem that particular courses (engineering and science courses at universities) are allegedly too theoretical. All three somewhat beg the question of the comparative advantage of the world of education in the promotion of theory. It may be useful to conceptualise their suggestions in terms of the ambits of education, training and production.

In Achi's paper, the theorisation of engineering, and the alleged unacceptability of engineering graduates are tackled by an educational solution: more engineering education to make graduates more acceptable to the world of industrial production. This seems like a "diploma disease" solution, but before condemning it, it is essential to understand the nature and culture of industry in Nigeria. Bringing the world of production into education in Nigeria is different from attempting to do the same in India because the technology of industry in Nigeria is not that deeply rooted.

In India, Kulkarni's experiment is an example of "dealing" with alleged university irrelevance by moving courses into the world of training, and finally into the world of production itself. The part-time training is highly specialised, very closely and explicitly related to the needs of the chemical industry, including their criticisms of the world of education. The ensuing course, therefore, is in the world of training, and is almost like in-

industry training, where the training is defined by the needs and technology of the particular industry. What this proves about the earlier undergraduate education courses is not at all clear. These are different worlds. In fact Kulkarni's group takes fresh chemistry graduates, who seem to be quite well prepared for his vocational-training-cum-production. Therefore the fact that the vocational training is successful may not have any implications for education. Not enough illustration is given in this example of the diversity of the industrial culture in India, and especially the very powerful attitudinal push towards self-employment.

In Sethu Rao's case study the experiment takes place in the world of education and the attempt is to introduce something of the world of external science research into the colleges. This is not a question of introducing production, but rather of research production. The students are put in touch with the research labs (CSIR, NAL, etc.) but it should be recognised that these labs find themselves in turn criticised for not being relevant to industry.

From Dr D. McCormick

Dr King has put forward an interesting analysis of the three papers. He is right to point out that the world of education is not the same as the world of industry. However, he is wrong in his belief that there are clear-cut demarcations between education and training. By his definition, much of what many of us in higher education teach cannot be considered to be education because it also features prominently in the education pro-grammes of industry, albeit under the co-ordination of training officers!

There is a second area in which I believe Dr King clouds the issue. Whilst he is right to insist that we take note of the cultural contexts of education, his argument fails to recognise the really important message which has come out of our present discussions. That message, which is relevant to India, to Africa, and to the rest of the world, is that there is room for innovations, in particular for innovations which actively engage students in investigating "real world" problems. This is not to say that all that we at present do in higher education is wrong. This is not the case, even quantum mechanics, which received a bad press from some speakers at plenary sessions at the conference, has its place at the heart of science teaching in the tertiary sector. It underpins much of modern science and is also highly relevant to technology. While too much of it in undergraduate courses is probably a bad thing, little discussion of it is equally a mistake.

Many of the papers point out omissions in our courses but they also show clearly how it is possible to rectify them without sacrificing our teaching of scientific principles.

Section I
Technical Training for Development

Introduction

Industry depends on technician training for the operation and maintenance of its resources, whether the industry is in a developed or a developing country. But the need is more pressing in developing countries. In the short term, such countries meet essential shortages by either adopting or adapting the expertise of developed countries or by importing technical help to maintain the capital equipment used in industry. In the long term, a shortage of technicians can only be met by developing good training courses (which in turn means developing good courses for trainers of technicians) and by increasing the value, esteem and rewards afforded by society on the work that a skilled technician does.

The term "technician" is used in different ways in different countries. This very important group of technical personnel comes in between the scientists and engineers on the one hand and the less skilled workers on the other. The responsibilities of technicians are mainly applying technology to a large range of field operations involving (a) production and construction, (b) testing and development, (c) installing and operating engineering plant, (d) estimating costs. The technician acts as liaison between the engineer and the skilled craftsman. Hence he has to interpret the engineer's plans and designs, indicate the production and construction techniques to be used, and choose the machines and tools best suited for the job. This means that he is responsible for a large number of professional and semi-professional functions which he has to carry out on *his own initiative* under the guidance and general supervision of a professional engineer or scientist.

F. O. Otieno of the Appropriate Technology Centre in Kenyatta College in Nairobi has pointed out some of the conflicts in technician training systems:

> "It has to be decided whether the training of industrial manpower is organised on a national basis or on an industry basis. There are advantages and disadvantages on each side. Training at national level usually avoids duplication of the training which may be common to many different industries, and it also simplifies administration. On the other hand if it is left to industry, it enlists loyalty to individual industries, it allows for individual differences between industries, and it also allows a measure of competition, which may result in improved training efficiency.
>
> For most developing countries, a compromise is best, some national

training and some industrial training. If the system is voluntary, it is possible only a few companies will co-operate. If there is legislation for compulsory training schemes, there might be disadvantages for the large countries if they had to do it and the smaller firms were excused. Yet another issue is whether the government or industry should pay for the training. It can be argued that government should pay because people are the nation's assets and investment in their training is one of a government's prime duties. However if industry does not contribute to the training, then it may take it for granted and lose interest. It can be argued that industry should pay for the benefits it receives. In the end, incentives may have to be introduced: this may be in the form of tax rebates."

The first contribution in this section from Iyengar discusses some of the problems of technician training in India. Saran and Balu also identify problems facing developing countries and the role which the Colombo Plan Staff College can play in alleviating some of these. The training of technicians on short courses on the maintenance and repair of scientific equipment used in tertiary education is the subject of the next contribution, in which is described a course arranged by ICSU-CTS at the Kenya Polytechnic at which academics and technicians became together better versed in techniques and strategies for the maintenance and repair of equipment which they use in their practical classes. The paper by Rivers refers to a CTS initiative to improve communication: CTS publishes an International Newsletter on Science Technician Training.

Meyer describes an interesting series of projects in Australia to help students gain experience of technical practices within post-secondary institutions. The final contribution to this section comes from Gadgil and Gambhir, who describe the Workers' University in India, which considers the needs of the least skilled.

During the workshop at the conference, Dr M. N. Rao, the professor-in-charge of the Technical Teachers' Training Institute on the S. J. Polytechnic Campus in Bangalore pointed out the difficulties faced by such institutions as his and he gave a description of the work done by the four Technician Teacher Training Institutes in India — in Bhopal, Calcutta, Chandigahr and Madras. The TTTI in Madras, for example, offers a long-term in-service course for teachers of technicians leading to qualifications (degrees and diplomas) as well as short courses on teaching methods, educational technology, curriculum development, educational management, student counselling and textbook preparation.

53

Technician Training for Development

B. R. N. IYENGAR

The Technician Education Society, Bangalore, India

The training of technicians and the development of their professional competence is a complicated process, to be done partly in a technical institute and partly in industry. The business of the technical institute is to give the prospective technician a sound broad-based knowledge in the theory of his chosen field of engineering. This has to be cross-fertilised with practical experience in industry so that technician trainees are familiar with working methods and skills relevant to their fields. All developing countries have to develop a system of technical education to meet their needs, so that industry can automatically accept those produced by the institutions.

Unfortunately this has not always been achieved and industry complains that the product from these institutions is unacceptable. Some industries expect or rather demand that polytechnics produce technicians equipped with adequate practical experience. It is inconceivable that any institution can achieve this without close co-operation with industry. The sandwich and block-release courses adopted in the UK are typical examples of such close co-operation. In the Federal Republic of Germany the pre-institutional apprenticeship exemplifies the participation of industry. In the USA organised programmes of apprenticeship are normal features, as are the "Technicoms" in the USSR. In developing countries there are several difficulties, particularly where industries are few and the demand for technicians is growing very rapidly. To cite an example, some years ago it was proposed to start here a post-diploma course in "Machine Tool Technology" by taking 15 to 20 candidates who had a 3 year general mechanical engineering diploma and give them a one year intensive course, closely integrated with a large public sector Machine Tool Industry. After considerable discussion and persuasion the industry agreed to take no more than two candidates for training because the industry had found that a larger number hampered their production schedules. It appeared that in spite of there being a will, there was no way.

Status of technicians

An All India Council for Technical Education (AICTE) was set up to expand facilities for technical education and training. But it was common to find that those who failed to secure admission to a college or institution of higher learning entered a technical institution like a polytechnic as a last and very unwelcome choice. This attitude is associated with social resistance to manual work and, even more important, to the poor status accorded to technicians. Only improved status and the recognition of the importance of technicians will divert bright young men and women to technician training courses, but this could play a vital role in the overall development of the country.

Standards of technician training

Unfortunately there is a considerable erosion in the standards of training. This is due in part to a very large increase in the number of institutions and also to the larger intake into older institutions without any increase in the facilities, such as space, equipment and staff. Moreover several institutions started by private agencies without any government support have sought the advice of the AICTE. Furthermore, there has been great difficulty in getting properly qualified staff for the large number of institutions and in many cases the staff–student ratio is inadequate.

Engineer–technician ratio

The ratio between the number of engineers and technicians is about 1:1.5. The AICTE recommended a ratio of 1:3 or 1:4. (The calculations have ignored the large number of persons working in industry, who have no formal training as technicians, but by virtue of their "practical experience" of several years have been doing the work of technicians.)

However things have not worked out as desired. Mainly due to political decisions, a very large number of engineering colleges have been sanctioned in several states over the past 5 to 6 years with very heavy intakes, while the number of polytechnics has not been increased in proportion. This has resulted in adverse ratios. In Karnataka, it is now about 1:1.2.

In India it is difficult to evolve a rational basis for determining the engineer-to-technician ratio due to a number of factors. Our industry is still in a developing stage and heavily dependent on imported know-how, capital equipment and managerial techniques, and very few industrial enterprises make any distinction between engineers and technicians when recruiting personnel. Supply and demand play the important part.

Even though technological self-sufficiency is proclaimed as our national goal, the efforts do not reflect our aspirations. Industry must invest more in research and development, taking a more progressive attitude in moving towards self-sufficiency and this will give our engineers opportunities to utilise their potential abilities. This will help in removing the unequal competition between graduates and technicians.

Many organisations do not have well defined personnel policies and very often advertisements call for "either graduates or diploma holders" for the same job thereby making no distinction between engineers and technicians. Their argument is that in any case the industry has to train them for their specific job.

There are also complications when a large number of persons with long practical experience, but no formal training, are to be promoted within the industry to the ranks of either engineer or technician. There are examples of good practice, however. One is at Hindustan Aircrafts Ltd in Bangalore, which asked one of the well established professional societies to conduct an impartial "diploma level" examination after jointly framing suitable syllabuses for them. A large number of employees have taken this examination over the years and reasonable justice has been done to senior employees with much experience, but no qualifications. There are other industries which have followed similar procedures.

The training of teachers of technicians

As in any educational institution, the heart of the training programme is the faculty. The AICTE has recommended a staff–student ratio of 1:12 for polytechnics. The present position in most polytechnics is much worse than this. However, it is the quality of the teacher which is more important and many polytechnic teachers lack industrial experience. There used to be great disparity in salaries between faculty members (both in engineering colleges and polytechnics) and their counterparts in industry. Hence it was very difficult to attract suitable persons as teachers in technical institutions. This gap has been narrowed in recent years.

To overcome the drawback of lack of industrial experience and knowledge, the Government of India has set up Technical Teacher Training Institutes (TTTIs) at several centres in India for polytechnic teachers. Initially the courses were for 18 months for graduate teachers, including 12 months' practical work in selected industries, and a 30 month course, including 12 months' industrial experience, for teachers with a diploma qualification. The period of training has since been reduced for long periods as there was no provision for substitutes on a temporary basis. Further, teachers were reluctant to leave their permanent homes for long periods, particularly as no incentives were given after return from training by way of salary increments or promotion.

If the programme of training teachers is to be effective many more such institutions have to be started as the number who can be trained in the present TTTIs is very small compared with the number of teachers to be trained.

The Government found another way to keep teachers updated in their own fields and financed a series of Quality Improvement programmes. This included sending teachers for higher studies and a scheme of "Summer and Winter School" programmes, which were on selected topics at polytechnics and colleges with the aid, initially, of experts from the USA. These courses have been very popular and well received by all concerned, but more such courses are needed for the large number of teachers.

Many well established professional bodies, for example the Institution of Engineers (India) and the Institution of Electronics and Telecommunication Engineers offer student membership and arrange continuing education courses, seminars and workshops. The tremendous increase in the enrolment of these Institutions for the studentship and associate membership is clear proof of the usefulness of these facilities.

54

Technician Training: The Role of the Colombo Plan Staff College

Y. SARAN and S. A. BALU

Indian Technical Teaching Training Institute, Bhopal, India
(Formerly, Director and Consultant at the Colombo Plan Staff College, Singapore)

Among various categories of manpower needed for development is the middle-level technical personnel commonly known as "technicians", who serve as a vital link between engineers and professionals on the one hand and the skilled workers on the other. Technician education systems, responsible for the production of the middle-level manpower, have responded to the national needs in varying degrees against the severe constraints within which they operate. While some countries have established systems for training of technicians others have only just started developing such systems.

The shortage of competent teachers, the low status of technicians and technician teachers, inadequate resources to match the increasing enrolment, rising costs, apathy of industry to share training responsibilities, difficulties in curriculum renewal to ensure relevance, and resistance to change and innovation are some of the major problems which technician education and training systems have to cope with. Yet they have to improve their effectiveness in terms of quality and quantity of technicians produced.

The Colombo Plan Staff College (CPSC) was created in direct response to the changing scene of the technician education systems of the regional countries facing these problems and issues. CPSC's main mission was to function as an instrument of regional co-operation and to assist broadly in the expansion and refinement of technician education and training within the diverse environments of the twenty regional member countries.

The changing regional scenario — problems and issues

The experiences of the Colombo Plan Staff College for Technician Education have led to the identification and appreciation of a variety of problems common to most of the regional countries.

Management of technician education

The major problems related to the management of the system are:

- absence of explicitly stated goals and policies
- absence of harmony between technician education provisions and manpower demand
- lack of effective evaluation of technician education and training system and institutions
- low level of co-operation between technician education systems and industries
- the status of technicians and hence the technician institutions is felt to be low and uninspiring

Curriculum design, implementation and evaluation

The following are some of the major problems related to the technician curriculum:

- difficulties in ensuring relevance
- the testing and evaluation of technician students are not adequate
- wastage in technician education

Staff

Competent staff in adequate numbers is a necessary pre-requisite for the effectiveness of education and training in any sphere. Most of the regional countries face serious shortages both in numbers and in desired competencies as far as teaching staff is concerned. The following represents some of the major problems in this area:

- ineffective policies in respect of staff recruitment, appraisal, reward and incentive systems
- inadequate facilities for initial teacher training
- inadequate facilities for continuous staff development

Resources

The following are the important problems in the area of resources:

- management of resources
- lack of appropriate and adequate teaching-learning resources.

The role of the Colombo Plan Staff College

Even on the face of these problems and constraints, many countries are initiating and introducing innovations, sometimes sporadic and sometimes planned. These innovations deal with various aspects of management, curricula, staff development and resources. In response to this broad scenario of the regional countries, the role of the CPSC has been and is to:

- provide, both at the College and in regional member countries, courses of further professional education and training for key persons responsible for the planning, development, administration and supervision of technician education and training, and technician teacher educators;
- conduct study conferences for senior administrators including directors of technical education, principals and other key personnel from education and industry, at which problems of technician education and training, including teacher education and training may be examined;
- assist regional countries and institutions to undertake projects in the field of staff and curriculum development, teaching-learning resource development and utilisation;
- promote, co-ordinate and undertake research into the special problems of technician education and training;
- advise and assist member countries in planning and developing their technician education and technician teacher education systems; and
- collect and disseminate information on technician education and training and technician teacher education and training.

Activities during the last decade

To fulfil these roles within the limited resources available to the College and at the same time to meet as far as possible the rising aspirations and expectations of regional countries, the College has produced college-based and in-country programmes. The themes of the programmes have been diverse. Some examples are: Technician education planning and development; Curriculum design, development and implementation; Student evaluation; Evaluation of technician institutions and programmes; Management of institutions; Development of learning resources; Developing skills in module writing; and Developing skills in small-scale research. Study Conferences have been held for senior administrators of technician education in regional countries to help the College to identify priority needs.

The first initiative in developmental projects was the modular project on In-Service Teacher Development. Four modules having 34 units have so far been published and are extensively used in the regional countries for teacher-training. The other two projects relate to the modules for developing skills in small-scale research and to the development of teaching-learning resources for use in technician education across the regional countries. In collaboration with international organisations like Unesco and the Asian Development Bank (ADB), a number of programmes have been organised at the CPSC in areas like Exchange of information on industrial training; Rural development; Planning, implementation and evaluation of educational projects, etc.

The regional countries, as well as agencies like ADB which provide financial assistance to projects, have welcomed the idea of the College providing consultancy services in technician education planning, curriculum design, staff development and the establishment of teacher training institutions. During the later part of the second 5-year phase, the College has provided such services on a substantial scale to those regional countries requesting such assistance. This has been made possible through the involvement of College faculty and resource persons, in both regional and non-regional countries, with whom the College has an effective rapport, established through its earlier programmes.

During the second 5-year phase, the College has made a start in promoting research capability in the region through modules for developing skills of small-scale research produced by the College. Collaborative research with associates in the regional countries has been carried out in areas like policy-related processes and student evaluation. Case studies on significant innovations in the region have also been written by identified experts in the region.

The first book published by the College is on "Aspects of Curriculum for Technician Education". A book on Management of Technician Education and another on Module writing are in the process of preparation at the time of writing of this paper.

Information collection and dissemination, identified as a very important activity, is being undertaken in a very limited way by the CPSC Library because of the non-availability of funds for a full-fledged information service.

Conclusion

The initiatives of regional countries individually and co-operatively with inputs provided by the Staff College, mainly in terms of development of key personnel, exchange of information and project-assistance have resulted in a variety of innovations. These relate to the establishment of curriculum development centres, technical teacher training institutions, media centres, testing services and item banks and production centres and projects for development of teaching-learning resources in regional countries. There is ample evidence of the co-operative working of the Staff College and the regional countries having made impact on national systems. The role of the Staff College is now changing to areas like policy formulation, technical teacher training, development of teaching-learning resources, management development, evaluation systems, promotion of collaborative action research and information dissemination, and increased in-country activity in all these areas. The review of the Staff College undertaken by an international team towards the end of the second 5-year phase has not only appreciated the effectiveness of the CPSC during the past decade, but also reinforced the College's changing role.

55

The Maintenance and Repair of Scientific Equipment used in Tertiary Education: An ICSU Initiative

This contribution is based on material submitted by G. CURRELL of *Bristol Polytechnic, UK;* R. M. MUNAVU, *University of Nairobi, Kenya;* D. G. RIVERS, *Huddersfield Polytechnic, UK;* and P. K. SRIVASTAVA, *University of Delhi, India*

Technicians in a non-industrialised society often do not have the opportunity to accumulate the skills relevant to modern scientific instrumentation in their daily work experience. There is, therefore, the temptation to expect the basic training courses, where they are available, for science technicians to provide not only education and training, but experience as well.

This difficult situation is compounded when it is realised that the expansion and up-grading of the education systems in developing countries over the last two decades have been largely at the level of teaching staff and above. Science technicians are not well paid and their role is not highly regarded.

The demands made on the skills and knowledge of science technicians in a developing country are, in fact, much greater that those experienced by their counterparts in an industrialised society. The majority of developing countries do not manufacture the many different scientific measuring instruments which are in common use in science teaching. The implications of this for the role of the science technician and for the effectiveness of training programmes are far-reaching.

Existing equipment is required to provide a useful life far in excess of that expected by the manufacturer as the foreign exchange needed to provide imported replacements is difficult to obtain. The need to improve the level of maintenance provided by the technical staff is clear as is the range of skills they are expected to possess. Unfortunately, in many cases, the technicians do not have the necessary information in the form of equipment manuals which would allow them to practise the skills they do possess. Equipment has often been purchased or supplied under aid agreements from many different sources. Manuals do not always arrive with the equipment, often being retained by the local supplying agent and occasionally the manuals are not available in a language that the technicians can read. The training programmes provided for the science technician should recognise these issues and place an appropriate emphasis on maintenance and related skills. The enhanced need for this aspect of training for technicians in developing countries has only recently been recognised.

This need has been reflected in the many international gatherings in the last decade which have had this or related issues as their central theme. For example, the 18th session of the Unesco General Conference in 1974 adopted no less than 100 recommendations in the area of technical and vocational education. This was followed 2 years later at the 19th session of the General Conference by a specific resolution urging action on the issues and problems of training technicians for the maintenance of scientific instruments. A subsequent series of seminars in Nairobi and Baghdad in 1979, Dakar in 1980 and Singapore in 1980 and 1981 took place to develop regional strategies and action plans. All of these initiatives took place with substantial support from Unesco.

The seminars should lead to curriculum change and these changes will bring about long-term improvements in training provision, but immediate action is necessary. Short-term changes can be brought about by the use of short intensive workshops offered as in-service retraining to experienced staff. ICSU-CTS has been active in support of this kind of approach and initiated a series of workshops, the first in Bombay and the second in Nairobi a year later.

These workshops were attempts to find ways of helping academic and technical staff in higher education develop an expertise which could then be applied to the general range of maintenance and repair problems which they face during their normal work.

The efficiency of any instrument is dependent on the day-to-day care it receives. This care is provided by the technicians in the laboratories and also the staff who use the instrument and depend on its reliability. Thus, one very important aspect of the workshops was to try to bridge the communication gap between those who use and those who look after scientific equipment. In the Nairobi workshop, institutions were therefore requested to nominate two members of staff, one academic and one technical. Twenty-four participants from thirteen institutions in the following nine countries attended and successfully completed the workshop: Ethiopia, Kenya, Lesotho, Malawi, Sudan, Tanzania, Uganda, Zambia and Zimbabwe.

Workshop philosophy

The ultimate objective of this type of workshop is for participants to achieve a more fundamental understanding of scientific equipment, thereby developing an ability to both identify and rectify faults and also to maintain the performance of equipment up to the level of its specifications.

Scientists and technicians have considerable experience in experimental techniques in a variety of disciplines but there are frequently problems in understanding the operation of scientific equipment, particularly as its complexity continues to increase. This understanding is often limited by a lack of effective communications in the subject — the "language" of instrumentation.

The practical activities in the workshop were designed to form a coherent sequence in the development of this "language" so that when complete instruments were introduced, the participants were aware of some of the underlying practical and conceptual skills necessary for this type of work.

In this respect the workshop was much more than a course, it was an attempt to quantify the type of training required (under realistic conditions) for the development of these "hidden" skills and their associated "language".

The practical procedures for maintenance (as opposed to repair) of scientific equipment are very dependent on the conditions to be found in individual laboratories. The workshop was an ideal forum to develop realistic approaches to maintenance through the sharing of experiences between the participants and in the context of a growing understanding of the instruments themselves.

Workshop structure

The workshop consisted largely of practical activities, the majority of these being directed towards a deeper understanding of scientific equipment. It is possible to identify five areas of education and training which contribute to this understanding.

(i)   Selected concepts in basic science and electronics;
(ii) Basic technical skills;
(iii) Specific concepts associated with scientific instrumentation;
(iv) Specific skills associated with scientific instrumentation;
(v)   The co-ordination of aspects (i)–(iv).

The experimental work was loosely divided into four sections: Basic electrical experiments; Instrumentation circuitry; Preventive maintenance and Complete instruments. Within these sections a range of activities emphasised and tested different factors of (i) to (v) above.

Locally-produced equipment

Within the structure outlined above, Dr Srivastava gave a course of talks and demonstrations on locally-produced equipment for science laboratories. He used, as examples, pH meters and conductance bridges developed in a group, based on the Departments of Chemistry and Physics of the University of Delhi, headed by Dr K. V. Sane and himself. The work has been carried out under the auspices of the International Union of Pure and Applied Chemistry, Committee on Teaching of Chemistry (IUPAC CTC) with the collaboration of the Division of Scientific Research and Higher Education of Unesco.

The group uses only materials which are manufactured within India and whilst this reduces the versatility of the equipment it strengthens the vital philosophy of self-reliance. The apparatus gives results that meet the requirements of most laboratory courses at college and university level. Workshop participants were pleased to be able to verify these claims experimentally.

If a teacher or technician assembles, or helps to assemble, scientific equipment himself, then he will become familiar with it and this will make maintenance much easier. He will have enhanced pride in the apparatus and the equipment will be cheaper.

Workshop evaluation

A questionnaire evaluating the content of the workshop was administered on the penultimate afternoon. The results were analysed and discussed with the participants on the final morning of the workshop. The evaluation showed that participants felt that the workshop objectives were highly relevant to their work situations. All of the elements of the workshop were seen to have been of considerable value and the balance of time allotted to them seemed to have satisfied participants. Responses showed clearly that the skills learnt during the workshop by participants had given then an increased confidence in their ability and willingness to develop and introduce planned maintenance programmes.

Overview of the workshop

The content of the workshop did not fall into the mainstream of either science or engineering courses and as such the exercises were found to be valid for all participants. The potential problems of piloting 24 participants with varied backgrounds through several sets of exercises with limited equipment in a very short time were not as great as might have been imagined. This is a clear indication of the very good level of support provided by the host institutions, the Kenya Polytechnic and the University of Nairobi, and also the flexibility and positive approach shown by participants and consultants alike.

It appears that after a two-week workshop of this type, the participants can expect to be able to:

(i)   Develop a working understanding of the fundamental elements of simple equipment such as temperature control units, pH meters, basic spectrophotometers;

(ii) Identify and test the power input and rectification stage of simple instruments;

(iii) Establish a basic maintenance record system;

(iv) Be receptive to further teaching material or information on scientific instrumentation.

56

Improving Communications Between Trainers of Technicians: A New Newsletter

D. G. RIVERS
Huddersfield Polytechnic, UK

The ICSU Committee on Teaching of Science (CTS) as part of their overall concern for the education and training of technicians contacted individual teacher trainers worldwide and asked them to identify problems they face in their work. One point raised many times was the lack of accurate and up-to-date information on training methods and courses which is readily available to those involved in curriculum development in other areas of education.

The International Newsletter on Science Technician Training is the result. It is concerned with all aspects of the initial and in-service education and training of science technicians, and therefore, addresses itself to the worldwide community of science technicians, the technician trainers and teacher trainers and the curriculum developers and Government officials who are responsible for science and technology.

In particular, the Newsletter acts as a forum for the presentation of new developments thereby contributing to the establishment of an enhanced technical environment with particular regard to the needs of the science technician.

Progress of the Newsletter

The first issue of the Newsletter was published in 1981 and the fifth appeared in 1985. It is therefore appropriate to consider how well the original aims are being met and what new developments are taking place.

A series of individual country reports on the education and training of science technicians in Algeria, Australia, Bahrain, Canada, Czechoslovakia, France, Ireland, Jordan, Seychelles, Tunisia, United States of America and Zambia have been published to date. Further reports from Netherlands, Nigeria, Sri Lanka, Federal Republic of Germany and Tanzania are available and will be included in future issues.

The process of sharing the information is only as effective as the quality of the mailing list allows it to be. The initial list was built on the information available at the Centre for International Technical Education at Huddersfield Polytechnic and from the members of ICSU CTS. This resulted in the first issue of the Newsletter being sent to 560 individuals in 75 countries. A steady stream of requests for additions to the list has resulted in these figures rising to over 850 in 103 countries. Whilst this is a satisfying increase, it is inevitable that many individuals and institutions who could usefully receive the Newsletter are not yet on the list.

The role of the Newsletter in the future

It was essential in the early issues for the Newsletter to establish its usefulness. The current level of response indicates that this has been partially achieved.

It is now necessary for the Newsletter to make a real contribution towards the solution of the problems outlined in this paper. It would be unwise to anticipate a set of general "solutions" with universal applicability. Nevertheless, although each country and often each institution has its own unique characteristics there are common features and initiatives which having met with a measure of success in one situation can provide stimulation and help to others. The Newsletter is one of the ways in which this type of information can be widely shared.

57

A Successful Innovation: Link Courses

G. R. MEYER

Macquarie University, Sydney, Australia

A recent trend in Australian secondary school education aims to ensure that the curriculum is as relevant as possible to the present and future needs of pupils. This trend is associated with a concern that pupils should stay at school, or at least in formal education, as long as practicable beyond the statutory school age of 15 years and that this opportunity be extended to all citizens irrespective of socio-economic grouping, sex, religion or ethnic background. The various state and territorial educational systems have responded in various ways. One of these is the development of "link courses" with post-secondary institutions. These are proving to be especially successful.

While various educational systems define link courses somewhat differently they are usually seen as part-time courses for secondary school pupils conducted by and within post-secondary (higher) educational institutions such as Colleges of Technical and Further Education (TAFE) or Colleges of Advanced Education (CAEs).

Some states, such as Queensland, take a broad approach to link courses defining them as:

> any programme that facilitates the expansion of educational oppor-
> tunities by co-operation between two or more educational institutions.
> These may be neighbouring schools; they may be a school and college
> of technical and further education, a college of advanced education, a
> pastoral or agricultural college, a special school or a field study centre.

Other States have more specific definitions. New South Wales, for example, defines them as follows:

> A link course is a co-operative programme for school students,
> involving the school and post-secondary educational institutions. Such
> a course will be developed and implemented jointly by the staffs of

both institutions and will be recognised as an integral part of the educational provisions of each.

In practice most link courses in Australia are arranged conjointly by secondary schools and Colleges of TAFE. Western Australia, for example, reports that "these courses have generally involved a small group of high school students attending a nearby technical college to take a course specifically designed for that group".

Usually the targeted group is for pupils aged 14 to 17 but some systems, notably the TAFE network in South Australia, limit the programme to pupils who are "at least fifteen years old, the age limit for compulsory school attendance".

Most school systems in Australia define the aims of link courses very broadly. The policy of the Education Department in New South Wales is typical: "the general aim of a link course is to extend the education of school students by increasing communication between the various sections of education, thus making the best uses of available resources".

The overall educational outcome from link courses is well summarised in the following statement by the New South Wales Department of Education.

"The general aim of a link course is to extend the education of school students by increasing communication between the various sectors of education, thus making the best use of available resources.

In particular a link course can contribute to school students' education by:

— providing more educational options than are available through one institution.
— motivating students in the traditional, academic subjects by assisting them to discover the relevance of these subjects for employment and further education.
— encouraging students to regard education as a continuing process.
— providing special learning opportunities for students with special abilities, disabilities and/or interests.
— increasing students' understanding of the structure and interrelationship among the various sectors of post-secondary education.
— enabling students to develop a better understanding of their capabilities and interests.
— developing student awareness of the range of opportunities in post-secondary education through first hand experience.
— breaking down sex-stereotyped attitudes to courses and occupations.
— providing students with the useful personal experience of studying in an adult environment.

— helping students to develop the appropriate attitudes which will assist their transition to employment and/or post-secondary education.
— increasing students' vocational awareness by providing information about occupations including the educational requirements for entry to these occupations and possible career opportunities within them.
— assisting students to gain greater insights into the skills and attitudes associated with various occupations.
— providing students with educational experiences which enrich the non-vocational aspects of life."

Examples

The variety and scope of link courses vary enormously from system to system determined by the need and resources available. The Department of Education of Western Australia suggests the following amongst others:

New opportunities for girls: to encourage girls to consider a variety of non-traditional occcupation areas, and to provide opportunities to meet successful role-models from a variety of trade areas (3 hr/week for 10 weeks).

Introduction to metal trades for country students: to encourage country students to experience a range of metal trades, and to help students to consider the issues and problems of "moving away" to study (a 3-day block in a city college).

Drafting: to provide technical drawing students with an opportunity to appreciate future study requirements (4 hr/week for 6 weeks).

Building trades: to provide an introduction to the building industry (6 hr/week for 11 weeks).

At the Royal Melbourne Institute of Technology in Victoria, there are courses (3 hr/week for 10 weeks) in Woodwork; Plumbing and Sheetmetal; Electricity; Fitting and Machining; Electronics; Motor Mechanics; Farm Hydraulics; Welding. These are for school students, grade 10. For grade 11 students, there are link courses on Building; Electronics; Fabrication techniques, Laboratory techniques. Most of these are 3 hr/week for a year.

Some link courses are intended only as "taster" experiences. For example, Seaforth College in New South Wales offered introductory "minicourses" of about 2–3 hr work in Secretarial Studies; Engineering Trades; Fashion; Automotive Trade, etc.

Organisation

Careful student preparation is essential. Those participating must understand the purpose of any proposed link course. In some instances students should be involved in planning and administering courses. Teachers should assess the readiness of students to undertake a course out of school and in particular should consider aspects such as social and emotional maturity and likely benefits to specific individuals. Organisation and management must be carefully planned — aspects such as timetables, transport and so on need to be considered in relation to the programmes and policies of the two institutions concerned. An example of instructions given to students of a Sydney high school is reproduced below.

STUDENT INFORMATION SHEET

THIS SET OF NOTES IS TO BE BROUGHT WITH YOU
EACH THURSDAY AFTERNOON OF THE LINK COURSE

WHERE IS IT ON?

The Link Course is on at Randwick Technical College, corner of Darley Road and Alison Road, Randwick.

HOW TO GET THERE!

At about 12.35 all students should assemble in the park opposite Pitt Street and wait for their teacher.

Buses (Number 17) will depart Pitt Street at 12.40 p.m. and arrive at the college where you will be escorted in groups to your correct class.

HOW TO GET BACK TO SCHOOL!

You will be dismissed from college at 2.55 p.m. Please go straight to the Darley Road bus stop and wait for your teacher. Buses (Number 17) will take you back to school.

WHAT GROUP AM I?

Year ten has been divided into 7 groups. Each group has a different number. Each week your group has a different class.

WHAT TO WEAR!

You will not be required to wear school uniform on the day of the Link Course. However, for your own safety Technical College requires you to wear at least the following:

- long sleeve shirt
- trousers
- shoes (Note — thongs, joggers or running shoes will not be permitted)

WHAT TO BRING!

Periods 1–4 Thursday morning will be normal lessons.

Suitable dress, a pen and paper is all that you will need for the Link Course. However, a folder is recommended because the technical college teachers will be handing out pamphlets at each class. You will need these because there will be follow up work in career education lessons.

HOW TO BEHAVE!
Any misbehaviour reported by the supervising school teachers or the technical college teacher will be dealt with.

ANY FURTHER QUESTIONS?
If in doubt about anything related to the Link Course please ask the supervising school teacher, the technical college teacher or the careers adviser. He/she will be at the technical college each Thursday afternoon.

Hope you enjoy the Link Course

Careers Adviser

Staff support in both institutions is essential and should be sought by those proposing a link course. Consultation between interested related departments of both institutions should be encouraged. Staff not directly involved, but who may be affected by timetabling changes and other aspects, must be thoroughly briefed. Where the link course aims to broaden vocational opportunities the Careers Adviser, School Counsellor and other appropriate staff should be involved in the planning and organisation. The design of the link course, its aims, structure and content needs to be co-operatively planned by the staffs of the two institutions. A course already operating within a post-secondary institution should only be chosen if the school regards it as the best possible way of achieving relevant goals for the school students. The support of the executive of each institution involved is required since provision of resources, organisational structure and continuing administrative support is essential.

Parents must be fully informed of all aspects of link course offerings and no change of direction should occur without parental consultation. Each course must be explained in terms of purpose, aims, content, activities and organisation and administration and parental permission must be obtained for students to study away from the school. The roles and responsibilities of the participating students, the staff members of each co-operating institution and the parents should be communicated to and understood by all parties.

Since schools serve the community it is important for the community to be informed about courses and programmes available at its school. Local employers, service organisations, governmental agencies, clubs, professional bodies and other community organisations all have a part to play, especially in a programme of link courses.

Evaluation

There is no doubt that link courses have been a remarkable success and represent a major growth area in Australian education. In 1983 in Victoria alone, technical colleges received a subsidy of nearly $200,000 dollars for link courses and of course schools contribute extensively from their own regular budgets. The technical colleges in Victoria anticipated "rapid growth of expenditure to at least one million dollars per annum over the next two or three years". This pattern is consistent with trends in other states.

The educational impact has been highly significant. An evaluative study commissioned by the Northern Territory Department of Education in 1982 showed that programmes in that territory had been well received by presenters and participants. Students were aware of the aims of the courses and considered that they had been met by the programme. They saw link courses as broadening experiences, giving them opportunities to maintain or learn skills which they could not achieve at school because of the constraints of the formal system of examinations.

Strong positive attitudes towards the programmes are shown by both staff and students involved in link courses in New South Wales. There is general agreement that major benefits include extension of educational opportunity, increased exposure to adult learning situations oriented to the work environment and the acquisition of useful skills and knowledge.

In conclusion, therefore, link courses are highly recommended for those school systems which have not as yet implemented this type of approach. They achieve highly significant educational outcomes through maximising the use of existing resources, and are therefore economical and highly cost effective. In situations where resources are critical factors in innovation, programmes of link courses can make a significant input on the quality of secondary school education, and help to give relevance and meaning to secondary schooling.

58

The Workers' University in India

A. GADGIL and V. G. GAMBHIR

*Adarsha Vidalaya, Goregoan, Bombay and the Centre for Science Education,
Tata Institute for Fundamental Research, Bombay, India*

As a result of deliberate policy to universalise education, the school system in India underwent rapid and enormous expansion during the last three decades. The massive efforts undertaken included making primary education free and compulsory, opening many new schools, centralising the production of textbooks, giving scholarships as well as help towards buying books.

In the last 25 years the number of primary schools rose from about 18,400 to over 570,000, with an enormous increase in enrolment in Standard I from 16 to nearly 69 million. However the spectre of the problem of the dropouts continues to haunt us. Nearly 80% of the pupils enrolled in Standard I drop out before reaching the secondary school leaving stage.

The system can, in no way, ignore these drop-outs, and, therefore, man, possible ways to convert them into a socially useful productive force were tried. 16–17-year-old school drop-outs in a village are unable to earn their livelihoods. However, if they can acquire, say, some working knowledge of the maintenance skills needed to repair irrigation pumps, not only can they earn their livelihoods, but they will have the respect of the community as well.

What is required for this conversion is not a thorough theoretical knowledge of the pump design, but a working knowledge of the common defects and the necessary skills to put them right. To meet this need, short term courses have been developed by many institutions: one such institution is Shramik Vidya Peeth (meaning, "Workers' University") and is run by the Central Government.

Though there are a number of Industrial Training Institutes (ITIs), spread throughout the state, they are unable to fulfil the demands of both individuals and industry. It is common for them to receive 2000, or even more, applications seeking admission to courses like turning and welding, for which limited places are available. Shramik Vidya Peeth provides an alternative solution. It not only provides technical courses like turning and welding but also courses on the repair of appliances (pressure cookers and mixers, for example, as used in the kitchen) and on screen printing and photography.

A small textile processing unit finds it difficult to employ a full-time textile designer. However, it is possible to employ one on a contract basis. A short term course in the Workers' University provides an economic solution, both to the industry and the individual. Moreover, the courses like those on tailoring, embroidery, preparing jams and pickles etc., prove useful to housewives and even in providing an opportunity for illiterates to start a cottage industry. Crafts and arts like batik are also included in the list of 140 different courses organised and implemented.

Although the Workers' University provides what is essentially informal training, these courses are very carefully designed and structured, making them almost wholly practical-based. Well-equipped workshops, a good library and qualified instructors are provided, and at minimal fees. A feature worth mentioning is the flexibility of the timetable of instruction. Secondary school students and those already employed elsewhere can attend the part-time courses at a time suitable to them. Special vacation courses are also organised. It is not surprising to find university graduates attending these courses to get practical experience. There is the example of those who graduated in a subject like chemistry, and then took a course at this institute and later became an instructor in the same institute. Such people are not rare. The fact that everyone coming out of the Workers' University finds immediate employment indicates the success of the approach.

Section J

Making Curricula Relevant for Industry: The Role of Teacher Training

Introduction

Ramachandra urged (see page 221) that teachers should be trained in terms of industrial and technological needs, and Somerville pointed out (see page 00) the chasm which sometimes exists between industrialists and teachers. Industrialists generally have a poor realisation of the capability of education to provide its needs, and educationalists who have not had the opportunity to gain a perspective of industry usually have a distorted view of industrial life.

In this section, a pre-service course for teachers in Zimbabwe which aims to introduce industrial issues is described by Steward and Towse. Kapiyo then describes how a pre-service course for teachers in Kenya deals with physics in terms of technologies which are relevant to the country, such as materials science and energy resources. An important aspect of such courses is the student project which has been referred to in many contributions. McCormick describes an in-service course for teachers in which a workshop devoted to problem-solving using technology is a core activity, and Gardner describes another in-service workshop which involved industrialists and tertiary and secondary teachers.

In the workshop at the conference, G. Tanuputra described interesting work in Indonesia where all the Teacher Training Institutes have courses known as "Community Service Programmes" for third year students, one year before they complete their training. The courses are preceded by one month in which speakers from agriculture, industry, health services, etc speak to the students, who are then sent to villages in the rural areas, helping people on problems related to these themes. Some examples are:

— building hygienic toilet facilities,
— simple ways of obtaining water fit for household use,
— building cheap houses fulfilling health requirements,
— producing food products from agricultural produce (for example, soya-sauce and soya-bean cakes from the soya beans cultivated by the farmers in the community),
— conservation of fruits by manufacturing fruit juices and syrups.

The students work together with local advisers in agriculture, health and local industries and their work is evaluated. It is found that this experience is a most valuable part of their training as teachers.

59

Industrial Issues in the Science Curriculum: Training Teachers in Zimbabwe

P. J. TOWSE
University of Zimbabwe

J. W. STEWARD
University of Papua New Guinea

The chemistry syllabus in the pre-service teacher training course leading to the B.Ed. degree at the University of Zimbabwe now includes a course on industrial chemistry equivalent to 20–25 hours. This course, which we believe to be the first of its kind in Africa, starts by examining in detail the broad issues involved in exploiting the natural wealth of Africa in general and Zimbabwe in particular. These issues are social, economic and political as well as chemical and technological. Consequently, discussions embrace topics such as economic growth and development, capital investment, cost effectiveness, the role of multinationals, political decision making, regional co-operation, manpower planning, energy requirements and transport just as much as they do the basic chemistry of the industrial processes.

The continent's main chemical industries are examined in detail. They are discussed on a country-by-country basis and fitted into a broad pattern of economic activity. Finally, issues such as the conservation of resources, energy consumption and pollution are analysed in national, regional, continental and universal terms.

With the second most highly developed industrial infra-structure in Africa, equivalent to an annual output of over US$2 billion, Zimbabwe features prominently in the course. Some of the reasons for this industrial development are explored — including, for example, UDI as a motivating force for self-sufficiency in pre-independence Zimbabwe.

It is no accident that this course follows closely the philosophy of our book *Chemical Technology in Africa*[1] for it is our firm belief that a proper understanding of industry requires more than a simple understanding of the chemical processes involved. In the final analysis, it is people and the quality of their lives which are important and so the course, like the book, has been seen as an "anthropology of chemistry" rather than a mere "anthology of chemistry".

This course is, however, only part of the whole industrial awareness scheme. Another important element arises from first-hand experience of industry. Although this part of the scheme is awaiting implementation, the idea is simple enough. Each student is expected to visit at least three of a number of firms which have agreed to participate in the scheme. They then write reports on these visits and, on the basis of their impressions, decide which of the three they would like to study in greater detail during a period of up to two weeks with the firm. They have to formulate a strategy for obtaining as comprehensive a view of its operation as possible. This strategy is to include a checklist of questions they would like answered during their stay. These questions should reveal that they have already thought in some depth about the issues raised by their first visit.

Their strategy is shown to a "resource person" working in the firm, someone who will act rather like an industrial tutor during their stay. In consultation with the student, the resource person will draw up a suitable programme for the stay, during which time the student will nominally be regarded as a regular employee, or trainee, with the firm. The student is expected to build up a carefully detailed picture of the various processes and issues within the firm, not merely the chemical issues but the human ones too.

The student is expected to write a detailed report on the visit, copies of the report being given to the resource person and to fellow students. The student is then asked to lead a seminar in which the key issues emerging from the report are discussed. This seminar, carefully planned in advance with the university tutor and the resource person, is chaired by the resource person. Afterwards, the other students are invited to reflect on the issues which have been discussed and to submit to the tutor and the resource person their considered comments on these issues. This gives the resource person the opportunity to judge the depth of their understanding of the industry and the quality of their perception of his firm's operations and problems.

At a second seminar, convened perhaps two or three weeks later, the resource person plays a more active role, responding to these considered comments and rounding out the views of his firm which the students have formed as a result of the earlier seminar. At best, this second seminar gives the resource person a chance to respond positively to suggestions put forward by the students, suggestions which sometimes can add new dimensions to the firm's operations and even throw up interesting possible solutions to its problems. At worst, it gives the firm a "right of reply", an important safeguard, not merely to mollify the firm and ensure its continued support of the scheme but also as a balance to any inaccuracies or prejudices which have been nurtured by the student who had worked there.

In countries where there are very few industries, there is the problem of year-on-year repetition in the same few industries and the possibility of subsequent plagiarism of the seminar reports.

The time devoted to this part of the scheme depends on the number of seminars, and hence on the number of students. In addition to the up to two weeks spent in industry, one needs to allow about an hour for each of the two student seminars. It should be acknowledged that there are only small numbers of students involved at the University of Zimbabwe at the moment. With much greater numbers, the time involved could prove excessive.

Yet a further element is possible, that of industrial decision-making through simulation exercises. One or two interesting simulations have been developed in Zimbabwe. Special simulations are those developed for the computer and allowing for more individual work. A few of those developed in Britain, for example CONTCT[2] and HABER[3] are used on our course in Zimbabwe, but it will be some time before we are able to develop examples of our own.

In-service training for established teachers

Since we are not aware of any in-service industrial chemistry programmes in Africa, the brief comments offered here are speculative, although they indicate what we hope will be introduced in Zimbabwe in the near future.

The course outlined above could be covered by a distance learning package, by an intensive vacation course, or by a suitable combination of the two. We favour the third alternative, a sort of Open University approach. Established teachers, especially those living far from high-population centres, are unlikely to have access to adequate library facilities and so the distance learning materials will have to be carefully structured. Such materials are more difficult to write than most people think, which probably accounts for the poor quality of so many distance learning programmes.

Although the industrial experience could, ideally, be offered all the year round, it seems more than likely that difficulties over the release of teachers during term time would force this into the vacations, at least for the time being.

Resources

We have already mentioned the paucity of materials in Africa and hence the rationale for the first of the resources, *Chemical Technology in Africa*.[1] There is a need to supplement this with files relating to issues of national importance. These files will contain cuttings from local newspapers and magazines, together with copies of extracts from suitable journals and such international news sources as *The Guardian, The Economist, Time* and *Newsweek.*

The seminar reports will be another resource available to the students, and indeed a wider audience of teachers. They will provide valuable background for use in schools.

Further co-operation with those in local industries could lead to the production of a series of booklets for use in schools, particularly for the pupils. A set of these is planned in Zimbabwe as a result of co-operation between the local section of the Royal Society of Chemistry and various industries. It is hoped also to create a set of tape/slide programmes for use in schools.

Also useful in Zimbabwe is the set of publications produced by the Ministry of Mines on different aspects of the country's metallurgical processes. These are valuable sources of background information, both for teachers and senior pupils.

An interesting decision-making simulation, the *Mupata Gorge Controversy*, was developed at Gweru Teachers' College[4] to examine critically the issues raised by the demand for more electrical power in Central Africa, and by a proposal to meet this demand through a hydroelectricity project based on the flooding of the lower Zambezi Valley and the creation of a second artificial lake like Lake Kariba. The students have to represent the various interest groups involved in the controversy and analyse the impact of the hydroelectricity project "not only on the ecology and wildlife on the area but also on such factors as the fishing industry of the country, employment opportunities and transport".

Although the simulation patently concentrates on environmental impact analysis, it does have a strong bias towards appropriate technology. More to the point, however, is the fact that since it has proved so successful further simulations are being developed within the University's B.Ed. chemistry syllabus. One, *Hwange* concerns the possible creation of a coal gasification plant based on the considerable reserves of coal in western Zimbabwe. Zimbabwe's dependence on oil imported through the pipeline from Beira in Mozambique and by rail and road from South Africa is subject to external political pressures and forms the basis for a scenario in which petrol self-sufficiency is sought, somewhat along the lines of the Sasolburg Project in South Africa. Another, *Uranium* is concerned with the exploitation of the uranium ore reserves in the Central African Republic and the transportation of the ore for processing in France.

In due course it is hoped to report more fully on these, on the industrial chemistry course and on the industrial experience component, for the whole strategy is, we believe, a useful blue-print for other parts of Africa and the rest of the Third World.

References

1. Steward, J. W. and Towse, P. J. *Chemical Technology in Africa*, Cambridge University Press, 1984.
2. CONTCT, Schools Council Computers in the Curriculum, Longman 1983.
3. HABER, Chelsea College Science Simulations, Edward Arnold 1982.
4. Luginbuhl, I. A method for developing an awareness of the conflicts between Western type development projects and resource conservation in Zimbabwe, a paper presented to the first international FASE Conference. "Is science education relevant to the school leaver in Africa?", Harare, Zimbabwe, 30 August–4 September 1982.

60

The Appropriate Technology Centre for Education and Research in Kenya

R. J. A. KAPIYO
Kenyatta University College, Kenya

The Appropriate Technology Centre (ATC) was established at Kenyatta University College (as described on page 53). The College is concerned with teacher training and one aim of the Centre is to introduce physics to students at the College in terms of technologies relevant to the needs of the country, for example, material science, energy resources and physics of the environment.

Students design and develop projects using locally available materials for specific rural needs. For example, the students at the Centre have designed, developed and tested cookstoves, jikos, with improved insulation and which therefore save fuel (charcoal). They have also worked on sisal cement technologies, on solar grain-dryers for use in stores, on biogas units and on water heaters.

The extent to which these technologies are disseminated and adapted by the teachers after being trained at ATC depends, in part, on their inclusion in the secondary science curriculum. As in other developing countries, Kenya suffers from limited resources — a shortage of trained teachers, overcrowded classrooms, lack of equipment — as well as examination pressures on the teachers and pupils alike.

In addition to its education and research function, the Centre aims to disseminate the developed technologies into rural communities and to encourage creative "do-it-yourself" activities within these communities. Hence, the Centre is also functioning as a non-formal education training unit.

61

Some Reflections on a Design-Technology Workshop

D. McCORMICK

Manchester Polytechnic, UK

One of the requirements of our M.Sc. course in Science Education is participation in a one-week Design-Technology workshop. The course is aimed at graduate science teachers with a minimum of 5 years' experience. It is a 3 year course, the taught component of which takes place during the first 2 years on two evenings a week, whilst the final year is mainly concerned with the writing of a dissertation. The workshop takes place at the end of the first year, and is the course's only full-time component.

Twenty students (two intakes) have now gone through the workshop, and whilst their comments on it have been generally favourable, there have been some which have been extravagant in their praise; comments such as ". . . the most refreshing educational experience ever. . .", ". . . taught me that I had some creative abilities . . .", and ". . . made me realise that much of science teaching lacks a vital dimension . . ." Since these responses were freely made by professional teachers about an activity which we felt to be fairly low-key and which was not examined (!), some reflections on the workshop and its effect on the participants are in order.

During the many discussions which led to the design of the course, it became generally accepted that the place of science in the secondary curriculum was problematic — that it was possible to articulate the relevant questions, but rarely to provide adequate answers, and that although we can map the course of science education from its origins to the present, its future nature was unclear. With this in mind, we designed a course that was in no way prescriptive, but rather one which hopefully would provide the teacher with the necessary knowledge and skills to contend successfully with the problems of science education, or at least to understand the origins of his frustrations.

During the first year, we discuss the nature of science and technology, their complex interrelationships and the relationship of both to science education. With regard to technology activity, where the overriding imperative is not the search for understanding but rather "to make things happen, to make things work", the relationship to science teaching cannot be easily discerned. Nonetheless, there is a feeling that since science education claims relevance to modern technological society, there is a need to articulate relationships and to explore possibilities. The Design Technology workshop responds to this feeling.

There are many different forms that a workshop could take. We wanted one which was a complete departure from the students' science specialisms, consequently ours has similarities to those planned for design students; however, it also draws heavily on the experience gained during creativity workshops at the Manchester Business School. We start by dividing into groups of four or five and playing games of the "Zin Obelisk" type[1]: each member of the group is given certain pieces of information on the obelisk and by exchanging this information amongst themselves, the group members have to complete tasks such as determining the day of the week on which the monument was completed. Some groups rapidly finish their task, others simply end up arguing. Clearly this is an opportunity to discuss group endeavours and, by means of self-assessment tests,[2] to reflect upon the roles individuals play in groups, their strengths and weaknesses, how strengths can be enhanced and weaknesses minimised, and how the best can be got out of a team.[3] It is generally agreed that we need to be more tolerant of one another.

The next team activity is entitled "Thinking Afresh". Here the teams do exhibit tolerance, but find difficulty in coming to terms with problems which have, in effect, been turned inside out. Watching people do a jigsaw, about which some of the normal assumptions would be misleading, is both amusing and edifying. At this stage we spend some time discussing the generation and handling of new ideas, both crucial to innovation.

After some self-assessment tests[4,5] in this area the groups discuss innovation and gain some experience in creativity techniques, such as brainstorming,[6] synectics,[7] and creative analysis.[9] Tudor Rickards, who advises on this part of the workshop, believes that "the techniques are not that important. They help to demonstrate to people that they can come up with more and better ideas. I see the techniques as muscle-building machines."[9]

The stage is now set for a problem-solving exercise. The task this year was to suggest ways in which a computer-based aid for budding teen-aged inventors[10] could be marketed within the education system. As in the previous workshop, the idea generation techniques worked well — dozens of ideas were generated. The groups were positive in their approach, willing to listen to and feed off one another; to say "yes, and. . .", rather than "No. . ." They were also willing to say silly things, something which science teachers do not find easy!

By now, almost half-way through the workshop, we were ready for a short design-and-build exercise. In 1984 the teachers were asked to crack open and eat a soft-boiled egg with one hand held behind their backs. The eggs were very soft, the egg-cups of an unstable design, and the results messy. They were then given four hours to design and build a device to enable a one-armed person to eat a soft-boiled egg. The participants were divided into two groups, both of which succeeded in their task. The first group was analytical, imaginative and systematic. Their device was novel. The second group had one idea — a heavy egg-cup with a low centre of gravity — which they built. Despite their success, we anticipated problems with this group, and were ultimately proved correct.

Following this exercise, we discussed open-ended problems, the situations from which problems emerge and the solutions to problems. A boat, a bridge, swimming, are all solutions to the problems which arise from the existence of a body of water. The situation we chose, both this year and last, for our major problem-solving exercise was the existence of old people. Old people have many difficulties, social, financial and psychological. We asked the two groups chosen for the previous exercise to reflect upon the matter, to identify problems associated with the physical disabilities of old people, to generate ideas for solutions, to design, build and test their proposed solutions, and to report on their endeavours to a final meeting of the workshop.

The discussions of the first group were lively and productive: many problem areas were identified and solutions proposed. The progress of the second group was slow. There was a marked reluctance by all to assume any sort of leadership role. They nonetheless did identify an interesting problem — the difficulty old people have in extracting light bulbs. The first group chose as their problem the difficulties encountered by the infirm in gaining access to stored articles. They worked on it with great enthusiasm. It was sometimes necessary to curtail their creativity. They ultimately decided on a rotating storage device. They were so happy with their design that they kept dreaming up uses for it — for supermarket storage and display, for assembly lines, and for seasonal indoor gardens! They made a model, and as might be expected from this derivative of the wheel, it worked. The second group made little progress. They had an idea for a "lightbulb gripper" but were unable to convert it into a lightbulb extractor. We intervened but to little avail. They finally, in desperation, constructed a device which required two able-bodied people to demonstrate the extraction of a light bulb; it had all the potential to be lethal! The two groups observed each other; the first became happier with their relative success, the second increasingly sullen with their apparent failure. More by accident than design we were given the opportunity to reinforce the lecture material on group problem-solving. After the final presentations, we engaged in a lengthy discussion.

What do teachers gain from the workshop? Obviously, the group problem-solving techniques are useful. Some teachers wish to introduce them to their colleagues; they complain that, too often, working parties of teachers degenerate into forums for argument. Others think that pupils could gain much from exposure to such techniques. However, the praise the workshop receives is not simply the result of having learned something about synectics. When highly qualified and experienced science teachers claim that the workshop gives them insights about their own creative capabilities, we must ask why, in all their previous educational experience, this has never happened. Can it really be true that all the years spent in science laboratories have not revealed their creative potential as much as has a few days spent designing some simple artefacts? This might indeed be the case. It is possible to have an education in science up to graduate level in which one simply follows instructions: in which one's personal input is almost negligible; and in which one's participation is best described as passive. This is probably what the teacher who commented that science education lacked a vital dimension had in mind.

We discussed this issue at the end of this year's workshop. We all managed to find excuses for our apparent omission. The natural sciences are demanding disciplines; the relationships and theories are not easily gained, and time is limited. Laboratories too are demanding. The craft skills of science do not come easily, and what we term experiments are meant to illustrate and exemplify; consequently exercises must work. In such situations, student input could, in more ways than one, be harmful! There is truth in all of this — but it is a distorted truth. If we do not realise this, we will take pride in the fact that our laboratory exercises are almost fool-proof (lab. manuals are marketed on this basis). However, we will hide from ourselves the fact that students can enter a laboratory to start work on an exercise they have never thought about — hence failing to see that, in science, thought precedes action — generally get the right results — hence failing to get those insights and ultimately that wisdom which grows out of failure. Perhaps most important of all, students never play any part in defining those problems — hence missing out the very essence of the science activity.

Some of the workshop teachers suggested that there could be great benefit in science teachers associating themselves with the Craft Design and Technology movement in secondary education. The idea has much to recommend it. The movement has wide and admirable aims. These they attempt to realise by creating "an environment where an individual can discover something of himself, his aptitudes, the relevance of his ideas to other people's ideas" and by teaching the student "to employ the general principles of his civilisation, this is, experiment, analysis and development, but of necessity in an enquiring and personal way, and not by merely reproducing past forms and successes.[11] We can all learn much from this philosophy.

Some science teachers, through project work, have already recognised its worth. A Manchester teacher involves her pupils in the design of aids for the disabled. The girls have had several notable successes, one of which, an automatic distress warning system, has been developed commercially.[12] The design is very simple. The pupils "lacked both the knowledge and the facilities" to do anything more sophisticated. There is a lesson here for those who believe that innovation has to wait until the student has reached degree level.

The trouble with project work is that it is usually offered as an optional extra, thus permitting either teacher or taught to opt out. Surely room can be made within the science curriculum for activities which are the very essence not only of technology but of science itself. Admittedly in order to embrace such activities some time spent on traditional ones would have to be sacrificed. Teachers in secondary education might find this difficult; we in higher education have no excuse for not making the attempt. The general and vocational benefits could be considerable.

We planned our course and its workshop component so as to give teachers some of the knowledge and skills necessary to contend with their problems. For all of us the lesson of the workshop is that science-teaching, at all levels, could benefit from the inclusion of activities which engage the pupils/students in the defining and solving of problems. The rewards for so doing could be high. During the last decade the war-cry of science educators has changed from "All for science" to "Science for all", the belief being that an education in science offers social and cultural benefits as well as vocational. I believe that we delude ourselves if we believe that science teaching in its present form can realise its social and cultural functions. If we wish to use science for the personal development of young people, they must, some of the time, be personally involved. We need to experiment here — a not unreasonable expectation from scientists.

References

1. Francis D. and Young D. *Improving Work Groups: A Practical Manual for team building*, California University Associates, 1979, pp. 147–151.
2. Mottram R. *Team Role Assessment Forms*, University of Manchester, mimeographed.
3. Mottram R. *Team-skills management: notes for users*, University of Manchester, mimeographed.
4. Jones L. J. *Barriers to Effective Problem Solving*, Manchester Business School, 1984.
5. Kolb D. A., Rubin I. and McIntyre D., *Organisational Psychology: an experiential Approach*, Prentice Hall, 1971.
6. Osborn A. *Applied Imagination*, C. Scribner and Sons, New York, 1957.
7. Gordon W. J. J. *Synectics — The Development of Creative Capacity*, Harper and Row, New York, 1961.
8. Rickards T. *Problem Solving through Creative Analysis*, Gower Press, Farnborough, 1979.
9. Rickards T. *International Management*, April 1980, p. 42.
10. Taking the Fear Out of Inventing, Greater Manchester County Press handout, 21 November 1983.
11. Hudson T. Quoted in *Design Education in Schools*, Aylward B. (Ed), Evans Brothers, 1973, p. 7.

12. Barrett M. *In a Girls' School*, Hidden Factors in Technological Change, Semper E. and Coggin P. (Eds), Pergamon Press 1976, pp. 196–199.

62
Industry Initiatives for Science and Mathematics Education

M. H. GARDNER
Lawrence Hall of Science, University of California, USA

A new programme designed to bring secondary education and industry into partnership in a manner that will result in the natural use of industrial and technological examples in science and mathematics teaching began in the San Francisco Bay area of California in the summer of 1985. Thirteen industries (for example Lockheed, Hewlett-Packard, IBM, AT&T) employed 41 teachers in their laboratories for eight weeks. The Lawrence Hall of Science, University of California, Berkeley, a unit dedicated to the improvement of science education, helped these teachers to convert their work experience into one of value and enrichment for their students. The industries are the initiators; Lawrence Hall staff are the facilitators; and the teachers are the translators and deliverers to the students in this programme.

During the summer work experience, teachers were given 10% of their time by industry to devote to development of ideas and materials for use in their classrooms. Some that immediately emerged were short teaching units, case studies, computer software, audio-visual presentations, planned visits for scientists and technologists in the classroom, student visits to the industries. Surplus industrial equipment was given to some schools for their laboratories.

The education components, planned to encourage and assist teachers in the development and use of their technological/industrial experience in the classroom, were under the direction of staff in the Lawrence Hall of Science. They included an initial orientation session for teachers and industrial mentors, a mid-term work session devoted to classroom use of ideas and experiences and the preparation of draft materials, and a wrap-up session at the end to consolidate their learnings and efforts. Follow-up sessions are subsequently held in November and April to encourage continued planning and use and to assess results and effectiveness of the summer experience. To enhance prestige, these teachers are being named IISME Teacher Fellows and they become the first members of an Academy of Fellows which will grow over the years to a substantial community of teachers with industrial/technological experiences.

The IISME programme, which appears to be very successful in its first year, will now grow to include as many as 200 teachers in 1986 and expand geographically to other areas of the West Coast of the United States.

To summarise, the major objectives of IISME are to:

1. Increase the number of science, mathematics, and engineering graduates by motivating more students to study mathematics and science in high school.

2. Improve the quality of high school education by improving the quality of science and mathematics teachers and improving the curriculum and instructional tools.

When asked what changes they would now make in their classrooms as a result of their IISME experience, teachers stressed:

Communication Skills

— clear and concise writing
— accurate documentation
— articulate, engaging presentations

Group Collaboration

— group problem solving
— interpersonal relations
— social skills

Work-Related Skills

— following instructions
— adherence to deadlines
— high quality work
— focus on projects

Learning How to Learn

— acquiring basic skills
— acceptance and use of criticism
— distinguishing between the 'one right answer' and creative problem solving

Section K
Co-operative Education

Introduction

In the first paper, Wilson describes some of the initiatives in co-operative education in the United States. This system, known as sandwich courses in the United Kingdom, is obviously excellent, but there are drawbacks. It depends on the availability of places for students in industry and in times of recession this is difficult. It also depends on very close collaboration between industry and the college or university to see that the work given to students is appropriate. In the second paper, Meyer writes about a scheme in Australia that enables school students to gain work experience in industry. Visits to industry also take place during the pre-service course for chemistry teachers at the University of Zimbabwe (p. 331).

63
Co-operative Education: A Means of Relating Work and Education

J. W. WILSON
Northeastern University, Boston, USA

Charles Kettering, long the head of research and development at the General Motors Corporation in Detroit, Michigan, used to liken traditional higher education to a butt weld, in which two pieces of metal are joined end to end, because the only contact between academic preparation and the workplace occurs at graduation. On the other hand, co-operative education, wherein students combine periods of study and productive work, is like a lap weld, in which the metals to be joined are overlapped and welded at several points. After drawing this imagery, Mr Kettering would observe that everyone knows a lap weld is stronger than a butt weld.

The essence of co-operative education is captured in this simile. It is a scheme which purposefully integrates productive and relevant work into students' college education. Because it provides multiple periods of employment alternated with classroom study, a bridge is established between what is learned in the classroom and what is required in the workplace. The result is a superior education.

Co-operative education was first initiated in the United States at the University of Cincinatti in 1906 by an engineering professor, Herman Schneider. Through this novel approach to education, he sought to solve two problems he had observed. First, he noted that many elements of most professions, and certainly engineering, could not be taught adequately in the classroom, but required practical experience for mastery. Second, he found that most students worked part-time during their college years simply because they needed to — most often at menial jobs, unrelated to their career goals. Through co-operative education he found a way to satisfy students' needs for practical, state-of-the-art experience and for financial assistance.

For a period of 50 years, co-operative education remained an interesting but relatively obscure scheme in American higher education; in 1960 there were only 65 programmes in the entire country. During that decade and continuing to the present, however, interest in it as a valuable approach to education increased dramatically and a good many institutions throughout the United States initiated programmes. Today more than 900 US colleges and universities operate programmes of co-operative education with over 170,000 undergraduate and over 4000 graduate students participating in virtually every field of study.

A number of forces contributed to this considerable growth in the application of co-operative education: a research report documenting the values of co-operative education[1] constituted the impetus for the formation of the National Commission for Co-operative Education, a non-profit educational agency, which had, and continues to have, the development and expansion of co-operative education in the United States as its central mission; during the late 1960s and early 1970s, responding to charges of irrelevancy in higher education, institutions searched for and instituted other forms of education, most of which incorporated off-campus projects, activities and work experience. In 1968 the federal government adopted a policy of encouraging the development of co-operative education by initiating a programme, which continues today, of awarding discretionary grants to institutions to initiate new programmes or to expand and strengthen existing ones. The fundamental reason, however, for the great growth of co-operative education is the fact that important values accrue to the participants—students, employers and institutions.

Benefits to students from co-operative education

Research and observation have demonstrated that it enhances the process of education in several ways. Consequently, significant values accrue to students who participate in this form of education.

The work experience helps students clarify and test their career goals. Students who have not made even tentative career choices use the co-op experience to explore, first-hand, career possibilities. Those students who have made a decision are able, based upon actual experience, to test their interest in and suitability for the career field.[2,3]

The co-op experience motivates students, leading them to increased persistence to graduation and to greater academic achievement. This is so because students see important connections between what they are studying and learning, and what will be expected of them after graduation. Their education is more relevant.[4,5,6]

The work experience contributes greatly to increased self-confidence. Many students in our institutions of higher education have grave doubts of their ability to achieve adequately. This lack of confidence very often disrupts the learning process and leads to the self-fulfilling prophecy of failure. Co-op students almost always have successful work experiences, which help them to see themselves in a new light — as competent persons capable of achieving their academic and career objectives. This too, is a part of the strength of co-operative education.[2,6]

Co-operative education students develop skills and attitudes essential to successful and satisfying adulthood. The reason is they are given adult tasks and are expected to perform them as adults and because they interact with adults as co-workers. One of the frequently cited difficulties in the development of young people in our society today is that they are shielded from the assumption of responsibility and, hence, are stifled in their development from youth to adulthood. Co-operative education helps to mature college youths into responsible adults.[1,6,7]

Co-operative education enhances the learning process in other ways too. Typically, students experience delay in getting feedback from their teachers on the adequacy of their learning efforts. This is most often not the case when they are on work assignment. Feedback on performance is quick and therefore useful in guiding next efforts. Co-operative work experience also provides students with opportunity to practise what they have learned. Practise is one of the most recognised but most ignored learning principles. Co-op makes excellent use of it.

Finally, students are paid for the work that they do. This has the obvious value of helping students, sometimes dramatically, to pay for their education. With the costs of education as high as they are, and the resources to pay them as uncertain as they are, co-operative education can make a significant contribution.

Benefits to employers

Among the several benefits which accrue to employers who participate in co-operative education research has documented two which are especially significant. First, students are a source of capable, productive and cost-efficient workers. What they may lack in initial experience is more than compensated for, by their basic ability and their eagerness to learn and to be a contributing part of the workforce. Several studies have shown that co-op students earn their way with employers.[3,8,9] Second, co-op students constitute an important pool of potential after graduation as employees. For most large corporations the opportunity to identify and recruit co-op students to become full-time employees is the principal motivation for their participation in co-operative education.[10] Studies show that nationally about 40% of all co-op students return to a co-op employer full time after graduation. That percentage increases to nearly 55% among graduates who worked a full year or longer for an employer as a co-op student.[3]

Benefits to colleges and universities

Institutions initiate programmes of co-operative education principally because they see it as a significant enhancement to their educational programmes and as consistent with the view that a vital function of higher education is to help students prepare for effective and rewarding citizenship. Additionally, there are three specific ways in which co-operative education helps colleges and universities.

It aids the recruitment of students. Opportunities for relevant, practical experience and for financial assistance are important considerations for students.[11]

It aids student retention. Research studies indicate that between 15 and 20% more co-op students persist in their college programmes to graduation than comparable non-co-op students.[4,5]

It helps keep the curriculum up to date. Through student feedback about their co-op assignments and through other work–education interactions spawned by the co-operative programme, the faculty have a valuable source of current information about current industry/business practice and technology.[1,12]

Conclusion

It is clear that co-operative education is a method of education and instruction of considerable merit. Important values accrue to all of the constituent participants — students, employers, colleges and universities. As each of these participants gains, however, the ultimate benefactor is the society of which it is a part. Societies and the quality of life they engender are very greatly shaped by the twin forces of work and education. Co-operative education purposefully links the academy and the workplace, enhancing the efforts of each to fulfil its individual and unique function and, consequently, producing better students and more productive workers. The outcome for society is to increase productivity and to raise the quality of life.

References

1. Wilson, J. W. and Lyons, E. H. *College Work-Study Programs: Appraisal and Report of the Study of Co-operative Education.* Harper and Brothers, New York, 1961.
2. Wilson, J. W. *Impact of Co-operative Education Upon Personal Development and Growth of Values: Final Report to the Braitmayer Foundation.* Northeastern University. Boston, Ma., 1974.
3. Frankel, Steven *et al. Co-operative Education: A National Assessment.* Applied Management Sciences. Silver Spring, Md., 1978.
4. Smith, H. Stuart. "The influence of participation in a co-op program on academic performance." *Journal of Co-operative Education,* November, 1965. pp. 7–20.
5. Lindenmeyer, Ray S. "A comparison study of the academic progress of the co-operative and the four year student." *Journal of Co-operative Education,* April, 1967. pp. 8–18.
6. Tyler, Ralph W. "The values of co-operative education from a pedagogical perspective." Paper presented at the Second World Conference on Co-operative Education. Boston, Ma., 1981.
7. Brown, Sylvia J. *Co-operative Education and Career Development: A Comparative Study of Alumni.* Northeastern University. Boston, Ma., 1976.
8. Arthur D. Little, Inc. *Documented Employer Benefits.* Cambridge, Ma., 1974.
9. Phillips, Jack J. "An employer evaluation of a co-operative education program." *Journal of Co-operative Education,* XIV, 2, 1978. pp. 104–120.

10. Weinstein, Dena and Wilson, J. W. "An employer description of a model co-operative education program." *Journal of Co-operative Education*, XX, 1, 1983. pp. 60–82.

11. Korngold, Alice and Dube, Paul. "An assessment model for co-operative education program planning, management and marketing." *Journal of Co-operative Education*, XIX, 1, 1982. pp. 70–83.

12. Sparrow, W. Keats. "Syllabus revision through co-operative education: Adapting courses to the 'Real World'." *Journal of Co-operative Education*, XVIII, 1, 1981. pp. 94–98.

64

Work Experience for Pupils in Formal Education: Principles and Practice

G. R. MEYER

Macquarie University, Sydney, Australia

In recent years there has been a worldwide trend towards developing stronger links between formal education and the world of work. Vocational education, in the broadest sense of that term, has had a long tradition of contributing to personal development and to general education, particularly in the secondary school. In most countries, however, perhaps with the exception of the United States of America which has demonstrated successful practice in this area since the turn of the century, direct work experience as part of a formal educational programme is a relatively new concept, especially at school.

The emphasis given to transition education has greatly accelerated in those countries with unacceptable levels of unemployment or with uncertain economies. It has been recognised, for example, that the school needs to take a more direct role in helping school leavers find suitable employment, and that this role should involve a great deal more than the older style careers advising. By providing an opportunity for work experience many formal educational systems have responded to this challenge creatively and effectively.

This paper attempts to describe a model for a work experience programme within a formal programme of education. While it refers specifically to current practice in Australia, this is only to provide an example of a programme in action. It is hoped that the model would have more general application.

Definition

The Department of Education in New South Wales defines work experience as "a tool or technique which can be used in the social development of students. It involves participation by students in work situations while they are still attending school. Its purpose is to assist students through the transition from school to the working life of the community."

Other educational systems open the approach to higher levels of education. The Tasmanian Department of Education states that "work experience is an integral part of a school or college's Participation and Equity Programme conducted for students enrolled either part-time or full-time. Such programmes involve those periods of participation . . . by students in work situations."

There is no reason in principle why the definition could not be extended to cover all levels of a formal programme of education. No doubt, however, the most critical level is the last year of compulsory schooling which in Australia is School Year Ten (secondary school form 4). Nevertheless work experience programmes are also important in senior secondary school, technical education, university and so on. At these levels, however, the objectives of work experience may be somewhat broader in aim, not only helping in transition from formal education to the working life of the community, but also in substantive areas of the curriculum such as a university course in industrial chemistry or a technical college programme in motor mechanics.

Aims of work experience

Most school systems in Australia limit work experience for each pupil to only one or two weeks, usually in School Year Ten. They do however build this in to a more extended programme of transition education and therefore tend to claim that the experience has the following types of aims.[1]

"— to break down the barriers between school and work
— to give students some knowledge of a variety of employment fields
— to give students an appreciation of the function and nature of work in our society
— to help students gain a greater knowledge of themselves enabling them to make a more realistic assessment of their abilities
— to provide new aspirations for the less motivated student
— to heighten awareness of the role education plays in future vocational pursuits
— to involve both parents and employers in the school to work transition"

The implication is that work experience breaks down barriers between life at school and life in the community at large. Barriers to this have been identified: (i) isolation from the reality of work; (ii) lack of skill by students in evaluating their personal strengths and weaknesses in relation to vocational decisions; (iii) low level of student confidence and social competence because of the isolation of the school environment; (iv) limited opportunities for parents, teachers, employers and students to interact about issues such as further education, employment and unemployment and (v) failure of the academic curriculum to prepare students for the work environment and to enter the labour force.

The extent to which work experience programmes help to overcome these barriers provides a measure of their effectiveness.

Organisation

Most work experience activities, including school based programmes in Australia, are organised in the following way.

Selection of Students. "Work experience programmes have relevance for all levels of ability and will have a different emphasis depending upon students' needs."[2] Ideally, therefore, all students of a particular educational institution should be involved. Where places are competitive however priority should be given to potential early leavers. Matching of students and jobs is usually done in consultation with other teachers, guidance officers and the administration; with students' home address, academic record and career preferences being important criteria. Most institutions establish waiting lists of students who can be placed as positions become available.

Selection of Employment. Most systems recommend that institutions keep a register of successful work experience placements. Criteria for selecting employment frequently include the following.[3] It is important that

"— the student is in no moral danger (situations of special concern include placing girls in factories where there are few other women or billeting students with farm workers);
— the student is not asked to carry out any work too difficult for his or her physical strength;
— the student is not exploited as cheap labour to do tasks with no educational value;
— the student is not occupying a position to the exclusion of a full-time worker;
— the student is given no responsibility beyond that which could be reasonably expected."

In addition it is desirable to select employment with some relevance to a substantive area of the curriculum. Most institutions mount public relations programmes to attract potential employers by stressing benefits to commerce and industry.

Types of Release.   Three types of release are commonly organised. These are: (i) continuous or rolling release where students attend work on a regular basis — such as every Monday afternoon over twelve or eighteen weeks; (ii) block release, the most frequent type, where students spend one or two weeks full time on a job and, (iii) an expanding programme aimed to help students who may be at risk by easing them into the work force over a period of time — starting say one day per week and gradually increasing the involvement to five days a week. The latter approach is a relatively recent development in Australia where it has been particularly successful in the schools of the Northern Territory.

Conditions of Employment.   Since work experience is part of formal education, students work without pay. They must, however, meet all industrial requirements, be approved by relevant industrial Unions, follow all safety rules, keep to formal working hours, and in effect work under the same conditions as regular employees. Insurance is generally taken out by the organising institution and all students are covered by the usual Workers Compensation Scheme. In some countries and states these aspects are regulated by formal Acts of Parliament or by ministerial regulation.

Preparation of Students and Parents.   Both students and parents need to be carefully briefed about the objectives and characteristics of work experience, and in particular about the roles of institution, student, parents and employer. Some systems, including for example the Education Department of Western Australia, have not only published relevant booklets and pamphlets to help in this preparation but also have produced films (or videos) and slide-tape materials.

Supervision.   This is a major undertaking, but in most systems each student on work experience should be visited at least once by a member of the educational institution. These visits should aim at overcoming organisational problems, maximising educational outcomes and establishing rapport with employers. In some cases where the placement is not satisfactory, alternative placements can be arranged by the supervisor. The Western Australian Department of Education recommends supervisors *at least* to address the following issues:

"1. Does the student enjoy the work experience?
2. Does the student appear to fit in?
3. Is the student dressed appropriately?
4. Does the employer and other employees seem satisfied with the student?
5. Is the workplace suitable?"

It is generally recommended that the visiting supervisor speak separately to both student and employer.

Student Reporting.   In order to maximise the educational impact of work experience most programmes require students to keep simple diaries. In Australia some States, of which Western Australia and Tasmania are good examples, publish such diaries and they are issued to students free-of-charge. These diaries require different aspects to be pursued each day such as (i) general description of the firm; (ii) responsibilities of the employees; (iii) interview hints; (iv) organisation of tasks or jobs; (v) safety and compensation issues; (vi) training and advancement; (vii) payment and Unions; (viii) other tasks seen but not experienced; (ix) educational requirements for careers in the work area experienced and (x) general opinions. Sometimes a page is provided for the employer to list further questions to be followed up by the student.

Follow-up.   Successful programmes build on the placement experience by involving students in discussion with teachers and with other students. Some require written reports which serve as group seminar papers. In addition wherever possible substantive content areas are pursued within the relevant subjects of the curriculum. A central aim of the follow-up activities is to help students identify their successes (and perhaps failures) while on work experience. Frequently, for example, students find they have weaknesses, say, in writing, in certain aspects of mathematics, or in substantive areas of science, and remedial work can be organised by the institution. Special discussion groups are frequently set up involving parents and sometimes representative employers may attend. Some institutions require all those involved to complete short questionnaires and use the information obtained (i) as input for discussion groups and (ii) to help in determining policy.

Resources.   Usually one member of staff of an institution is required virtually full-time as a work experience co-ordinator. His or her main responsibility is to organise placements. All members of the teaching staff, however, need to be involved in placement policy; in the preparation of students and parents; in supervisory visits and in follow-up activities. There is need for extensive clerical and administrative back-up and resources must be provided for liaison with parents, with employers and with the community at large. Public relations documents and other resources are required and must be effectively distributed. The Tasmanian Department of Education, for example, widely distributes pamphlets such as the diary already mentioned above, and information for parents; guidelines for employers; guidelines for Trade Unions and safety hints for students.

Outcomes

Work Experience activities involve a great deal more than traditional vocational guidance and much more even than "job sampling". When viewed as part of a programme of transition education they contribute to the personal development of students. Seen in this way they perhaps may be better designated as programmes of "Work Education". *Work education* should provide insights of an economic, social and political nature and students should use the work experience opportunity perhaps as a case study — part of a broader analysis of society and the world of work. In the words of Norman Curry, The Director General of Education in Victoria, "In secondary schools, work experience is not aimed at giving students specific vocational skills but rather is intended to broaden their understanding of themselves and the work environment".

There is, however, another important aspect. If properly organised, work education can also contribute in a very real way to the substantive content of the curriculum. In science, for example, biology students can experience work on farms, in horticulture and in biologically based secondary industry. Physics students can be placed in electronic factories, in medical facilities, in computing and so on. Chemistry students can work within chemical laboratories, in pharmaceutics, in chemically-based manufacturing industries and in similar jobs. Geology students can find work in mining, in certain aspects of engineering and so on. Such an approach of course makes placement more difficult — in certain districts suitable placements may not always be optimal — and the full co-operation of the teaching staff is required for such an approach. Nevertheless this "extended use" of work experience leads to much more than mere cognitive gain. Students appreciate the importance of a sound education for effective work. They perceive the relevance of formal studies and are much more likely to wish to continue to learn.

Conclusion

It would be unrealistic to claim that one or two weeks of work experience alone could solve all the problems of transition from formal education to life in the community at large. Seen, however, as part of a broader programme of personal development and as an extension of the mainstream curriculum, gains from work experience can be substantial.

In Australia, work experience programmes have been well received by students, parents and employers. In Victoria alone in 1982 some 60,000 students from both government and non-government secondary schools participated in one or more experiences of working in the community.

A recent evaluative study in Australia by Ralph Straton and Maxine Murray suggests that participation in work experience programmes has enhanced the social development of some students; has engendered more realistic attitudes to the workplace and has provided useful information about the specifics of certain types of employment.[4] There are, however, obvious dangers. Unless work experience is seen as part of a broad picture of personal development, including work education, and involving linkage to the mainstream academic programme, then there could be undesirable socialisation, job stereotyping and curriculum disjunction. An effective work experience programme requires the total commitment of an educational institution and involves a sensitive understanding of the personal needs of individual students.

References

1. Cole, Peter "Work Experience a Critique of Current Practices and Directions for Program Development" *Victorian Institute of Secondary Education Advisory Services and Guidance Branch Occasional Paper No. 1.* January 1981. p. 4.
2. Australian Capital Territory Schools Authority Work Experience Committee *First Annual Report 1981.* Canberra: ACT Schools Authority, 1982. p. 2.
3. Education Department of Western Australia *Work Experience a Guide for Teachers in Implementing Work Experience in Secondary Schools.* Perth: Education Department of Western Australia, 1980. p. 13.
4. Straton, Ralph G. and Murray, Maxine *Department of Education and Youth Affairs Work Experience in Secondary Schools.* Canberra: Australian Government Publishing Service, 1984.

Index

363